Minerals of Oregon

by Graham John Mitchell

with an introduction by Kerby Jackson

Introduction

It has been years since DOGAMI released this important publication "Minerals of Oregon". First released in 1916, this important volume has now been out of print for years and has been unavailable to the mining community since those days, with the exception of expensive original collector's copies and poorly produced digital editions.

It has often been said that *"gold is where you find it"*, but even beginning prospectors understand that their chances for finding something of value in the earth or in the streams of the Golden West are dramatically increased by going back to those places where gold and other minerals were once mined by our forerunners. Despite this, much of the contemporary information on local mining history that is currently available is mostly a result of mere local folklore and persistent rumors of major strikes, the details and facts of which, have long been distorted. Long gone are the old timers and with them, the days of first hand knowledge of the mines of the area and how they operated. Also long gone are most of their notes, their assay reports, their mine maps and personal scrapbooks, along with most of the surveys and reports that were performed for them by private and government geologists. Even published books such as this one are often retired to the local landfill or backyard burn pile by the descendents of those old timers and disappear at an alarming rate. Despite the fact that we live in the so-called "Information Age" where information is supposedly only the push of a button on a keyboard away, true insight into mining properties remains illusive and hard to come by, even to those of us who seek out this sort of information as if our lives depend upon it. Without this type of information readily available to the average independent miner, there is little hope that our metal mining industry will ever recover.

This important volume and others like it, are being presented in their entirety again, in the hope that the average prospector will no longer stumble through the overgrown hills and the tailing strewn creeks without being well informed enough to have a chance to succeed at his ventures.

Kerby Jackson
Josephine County, Oregon
May 2016

3

CLASSIFICATION OF OREGON MINERALS

Elements:

A. Non-metals.

Graphite	C

B. Semi-metals

Antimony	Sb

C. Metals

Copper	Cu
Gold	Au
Iridosmine	(Ir Os), sometimes with Rh, Pt, etc.
Iron	Fe
Josephinite	$Fe_2 Ni_3$
Lead	Pb
Platinum	Pt.
Silver	Ag

Sulphides, Arsenides and Tellurides:

Argentite	$Ag_2 S$
Arsenopyrite	Fe As S
Chalcocite	$Cu_2 S$
Cinnabar	Hg S
Cobaltite	Co As S
Galena	Pb S
Hessite	$(Au\ Ag)_2 Te$
Marcasite	$Fe S_2$
Molybdenite	$Mo S_2$
Pyrite	$Fe S_2$
Pyrrhotite	$Fe_n S_{n+1}$
Sphalerite	Zn S
Stibnite	$Sb_2 S_3$
Sylvanite	$(Au\ Ag) Te_2$

Sulpho-salts:

Bornite	$Cu_5 Fe S_4$
Chalcopyrite	$Cu Fe S_2$
Freibergite	(a silver bearing tetrahedrite)
Proustite	$Ag_3 As S_3$
Pyrargyrite	$Ag_3 Sb S_3$
Tetrahedrite	$Cu_3 Sb_2 S_7$

Haloids:

Halite	$Na\ Cl$

Oxides:

Agate	$Si\ O_2$
Chalcedony	$Si\ O_2$
Corundum	$Al_2\ O_3$
Cuprite	$Cu_2\ O$
Diatomaceous Earth	(Siliceous remains of microscopical plants or animals)
Hematite	$Fe_2\ O_3$
Ilmenite	$(Fe\ Ti)_2\ O_3$
Infusorial Earth	(see Diatomaceous **Earth**)
Pyrolusite	$Mn\ O_2$
Quartz	$Si\ O_2$
Rutile	$Ti\ O_2$
Sapphire	$Al_2\ O_3$
Tenorite	$Cu\ O$
Wad	(for composition see Wad)

Aluminates, Ferrites, etc.:

Chromite	$Fe\ Cr_2\ O_4$
Magnetite	$Fe_3\ O_4$

Hydroxides:

Limonite	$Fe_2\ (OH)_6\ Fe_2\ O_3$
Opal	$Si\ O_2 + n\ H_2\ O$

Carbonates:

Aragonite	$Ca\ CO_3$
Azurite	$Cu_3\ (OH)_2\ (CO_3)_2$
Calcite	$Ca\ CO_3$
Cerussite	$Pb\ CO_3$
Dolomite	$Ca\ Mg\ (CO_3)_2$
Malachite	$Cu_2\ (OH)_2\ CO_3$
Natron	$Na_2 Co_3 + 10\ H_2O$
Rhodochrosite	$Mn\ CO_3$
Siderite	$Fe\ CO_3$

Phosphates, Arsenates, etc.:

Monazite	$(Ce\ La,\ Di)\ PO_4$

Nitrates and Borates:

Borax	$Na_2\ B_4\ O_7.10\ H_2O$
Colemanite	$Ca_2\ B_6O_{11} + 5\ H_2O$
Soda Nitre	$Na\ NO_3$

Sulphates:

Alabaster	$Ca\ SO_4 + 2\ H_2O$
Barite	$Ba\ SO_4$
Celestite	$Sr\ SO_4$
Gypsum	$Ca\ SO_4 + 2\ H_2O$
Mirabilite	$Na_2SO_4 + 10\ H_2O$

Silicates:

Actinolite	$Ca\ (Mg,\ Fe)_3\ (Si\ O_3)\ 4$
Asbestos	$H_4\ Mg_3\ Si_2\ O_9$
Biotite	$(H,\ K)_2\ (Mg,\ Fe)\ Al_2\ (Si.\ O_4)_3$
Chlorite	$H_8\ (Mg,\ Fe)_5\ Al_2\ (Si\ O_6)_3$
Chrysocolla	$Cu\ Si\ O_3.2\ H_2O$
Chrysotile	$H_4\ Mg_3\ Si_2\ O_9$
Epidote	$H\ Ca_2\ (Al,\ Fe)_3\ Si_3\ O_{13}$
Erionite	$H_2Si_6Al_2\ Ca\ K_2\ Na_2O_{17} + 5\ H_2O$
Garnet	$(Ca,\ Mg,\ Fe,\ Mn)_3\ (Al,\ Fe,\ Mn,\ Cr,\ Ti)_2\ (Si\ O_4)_3$
Garnierite	$H_2(Ni,\ Mg)\ Si\ O_4.H_2O$
Heulandite	$H_4\ Ca\ Al_2\ (Si\ O_3)_6 + 3\ H_2O$
Kaolinite	$H_4Al_2Si_2O_9$
Muscovite	$H_2(K.\ Na)\ Al_3\ (Si\ O_4)_3$
Olivine	$(Mg,\ Fe)_2\ Si\ O_4$
Pyroxene (Augite)	$Ca\ Mg\ Al\ Fe\ (Si\ O_3)_2$
Rhodonite	$Mn\ Si\ O_3$ with replacement by Fe, Zn or Ca
Talc.	$H_2\ Mg_3\ Si_4\ O_{12}$
Thomsonite	$(Ca,\ Na_2)_2\ Al_4 \cdot (SiO_4)_4 + 5H_2O$
Zircon	$Zr\ Si\ O_4$
Zoisite	$H\ Ca_2\ Al_3\ (Si\ O_4)_3$

Hydrocarbons:

Asphaltum	(Mixture of different hydrocarbons)

SUGGESTIONS FOR MINERAL IDENTIFICATION

Form of Occurrence. If the mineral occurs in crystals, the determination of the crystal form will be an aid in identifying the specimen. For those not familiar with Crystallography, the reference to "Moses and Parsons" in the reference list will help make this point clear. In fact, any book on Crystallography will be of assistance.

The mineral may occur in compact masses, i. e., made up of fragments of crystals bound firmly together. When these fragments are large the term "coarse grained" is used; when small, "fine grained."

Other terms used in dsecribing the form in which the mineral occurs are: fibrous, lamellar, micaceous, columnar, bladed, nodular, granular, mammillary, botryoidal, reniform, oolitic, stalactitic, dendritic, concretionary and pisolitic.

Fibrous means that the mineral is made up of fibers or thread-like masses. Chrysotile (asbestos) is a good example.

Lamellar refers to a form built up in plates. Gypsum is an example.

Micaceous describes minerals which separate into very thin sheets like mica.

Bladed refers to crystals resembling a knife blade, i. e., long and thin. Cyanite has a bladed form.

Rounded lumps are termed **nodular.** Pyrite, sometimes, illustrates nodular form.

Granular refers to fine grained compact masses. Steel galena is a good example.

Mammillary refers to low rounded lumps of which Smithsonite is an example.

Botryoidal from the Greek meaning a bunch of grapes. Chalcedony illustrates this form.

Reniform or kidney-shaped forms are sometimes exhibited by hematite; also cinnabar.

Oolitic refers to forms resembling fish roe. The Clinton iron ore of Eastern United States illustrates oolitic.

Stalactitic forms resemble icicles in shape and are common in caves. Calcite is often found in this form.

Dendritic. Minerals which have a branching or treelike form are called dendritic. Native copper is at times dendritic.

Concretions are rounded masses formed by the accumulation of mineral matter around a central portion called a nucleus. Siderite is an example.

Pisolitic forms resemble shot, being larger than oolites. Bauxite is a good example.

The term **earthy** refers to a form resembling earth. Wad and Kaolinite are examples.

Hardness. (H) Hardness is stated in terms of Moh's scale of hardness, which is given in the glossary of this bulletin. To determine the hardness, try to scratch the mineral with the finger nail. If it is scratched, then the hardness of the substance tested is less than 2½ in the scale. The next thing to do is to take the mineral which is the next below in the scale, gypsum "2." By using a sharp corner of this mineral, gypsum, try to scratch the one to be tested. If "2" scratches the mineral, then try to scratch "2" with the unknown mineral. If the unknown scratches "2," then the unknown mineral and the mineral with the hardness of "2" have the same hardness; i. e., minerals of the same hardness scratch each other.

The hardness of any mineral can be determined by following the method outlined above. First trying to scratch the unknown with the mineral from the scale, then the mineral from the scale by the unknown. If a mineral is scratched by "4," say, and not by "3," then the hardness of the unknown is 3½, provided the unknown does not scratch "4." This is as close as the determination can be made.

Another means of determining the approximate hardness is to try scratching the mineral with a knife point. The ordinary knife will scratch all minerals having a hardness of 5½ or less. The minerals composing Moh's scale of hardness are readily obtained, except diamond, which is not necessary in determinig the common minerals.

Specific Gravity. The specific gravity can be roughly determined by hefting the mineral in the hand. Minerals, such as barite and cinnabar, are characterized by their noticeable weight. The Jolly balance is used in most laboratories in making specific gravity determinations. Illustrations of this instrument can be seen in most books on mineralogy.

Color. The color is a feature readily determined and is characteristic of such minerals as azurite, garnierite, realgar, etc.

Streak. The streak is the color of the powdered mineral and can be readily tested by rubbing the mineral on unglazed porcelain or by scratching with a knife-point until a powder is produced.

Cleavage. The ability of some minerals to break or cleave along more or less smooth surfaces more easily than along others. This cleavage is parallel to one or more of the faces of the forms in which the mineral can crystallize. In the case of galena, the cleavage in three directions at right angles is characteristic of that mineral.

Blowpipe Tests. Apparatus for making blowpipe analyses is not often found outside of laboratories or assaying offices. Arrangements are made to make such tests in most of the assay offices located at mines. The layman can, however, make a few simple tests, such as fusibility, in the case of the more fusible minerals such as stibnite, asphaltum, sulphur, etc., by holding them in a lamp flame. If he has access to the acids he can make solubility tests. Citric acid in solid form may be purchased at drug stores and a solution made by dissolving in water. This is useful in distinguishing carbonates. Where apparatus is not available and a determination of the mineral is desired, it is best to forward a sample to some of the local establishments where such identifications are made. Assaying offices, or the Departments of Geology in the State Universities or Schools of Mines, do such work.

In sending material, it is desirable to give notes on the occurrence; i. e., whether the mineral occurs in a vein, bed, placer gravel, etc. Also the locality, as near as possible, to the larger settlements or streams. Locations with respect to the section, township and range are the best. The sample sent for determination should contain, if possible, about four cubic inches.

NOTE.—The policies in vogue in the universities and schools of mines in the different states vary; however, the custom is not to compete with local assaying establishments, but rather to make qualitative tests and to give advice as to whether or not further investigation is warranted.

MINERALS OF OREGON
(Arranged alphabetically)

ACTINOLITE Ca (Mg, Fe)$_3$ (Si O$_3$) 4

Locality: **Douglas County, Roseburg** quadrangle, near Roseburg.

Distinguishing Features: Occurs in long prismatic monoclinic crystals which are sometimes arranged in radiating bunches. Actinolite also occurs in columnar and fibrous aggregates. Color, pale green to dark green. Hardness = 5½. Specific gravity = 3.0. Fusible at 4 to a dark glass. Insoluble in acids.

Occurrence: Occurs in talc and chlorite schists and in contact zones associated with talc, serpentine, and epidote. Common constituent of some igneous rocks.

AGATE
(See Chalcedony)

ALABASTER
(See Gypsum)

ANTIMONY
(See Stibnite)

ARAGONITE Ca CO$_3$

Locality: **Grant County, Dayville,** on South Fork of John Day River. **Josephine County,** Oregon Caves.

Distinguishing Features: Occurs in prismatic or acicular Orthorhombic crystals; also in columnar and fibrous masses and in incrusting and stalactitic forms. Twin crystals of the pseudo-hexagonal form are found. Hardness = 3½. Specific gravity = ± 2.9. Color, colorless, white or amber. Effervesces in cold or hot acid. Aragonite and calcite are distinguished by boiling the powdered mineral in cobalt nitrate. The powder will turn purple if aragonite, and remain white, with a greenish ring on the outer edge if calcite.

Occurrence: As a secondary mineral in seams and cavities of limestone where it is associated with calcite. In basic igneous rocks such as basalt, associated with zeolites. With gypsum in clays and marls. In hot spring deposits with sulphur. In mollusc shells as a lining called mother-of-pearl. In serpentines as a secondary mineral. In iron deposits of limonite and siderite.

Use: Source of lime.

ARGENTITE Ag₂ S
(Silver Glance)

Locality: **Baker County,** mined in Cable Cove and Rye Valley districts.

Distinguishing Features: Occurs massive, incrusting; also in cubical crystals. Hardness $= 2\frac{1}{2}$. Specific gravity $= \pm 7.3$. Color is dark lead gray to almost black. Metallic luster. Very sectile (easily cut with a knife). Under blowpipe flame fuses easily at $1\frac{1}{2}$ yielding on charcoal a malleable button of silver. Soluble in HNO_3 with the separation of S. HCL gives a white precipitate of AgCl, soluble in NH_4OH.

Occurrence: In veins associated with other silver bearing minerals.

Uses: Important ore of silver. "Silver Glance" is the miner's term for this mineral.

ARSENOPYRITE Fe As S
(Arsenical Pyrite)

Locality: **Grant County,** Susanville district. **Josephine County,** Maid of the Mist and Braden mine; Grants Pass district and Silent Friend mine. **Baker County,** mined with gold ore in Granite, Sparta, Sumpter, Cable Cove, and Elkhorn districts. **Jackson County,** Upper Applegate and Gold Hill districts.

Distinguishing Features: Arsenopyrite occurs in compact masses; also in well-formed crystals. The crystals are orthorhombic, similar in habit and angle to marcasite. Hardness $= 5\frac{1}{2}$—6. Specific gravity $= \pm 6.0$. Color is tin-white to light steel-gray with metallic luster. Fuses on charcoal rather easily to a magnetic globule and gives off arsin (garlic odor). When gently heated in a closed tube gives a red sublimate (As S); on further heating, an arsenic mirror is formed. Soluble in HNO_3 with separation of S.

Occurrence: As a vein mineral. (Mother lode, California, contains arsenopyrite in the quartz veins.)

Uses: Source of the white arsenic (As_2O_3) of commerce. Arsenopyrite is mined as gold ore at Deloro, Canada.

ASBESTOS
(See Chrysotile)

ASPHALTUM
(Mixture of different Hydrocarbons)
(Asphalt, Mineral Pitch, Bitumen or Mineral Tar)

Locality: Harney County, near Lake Alvord and near Drewsey, **Coos County,** in the coal of Coos Bay. **Wheeler County,** three miles east of Clarno in geodes. Two miles southwest of this locality it occurs as disseminated pieces (about the size of a pea) in the Clarno Eocene (tuff) formation.

Distinguishing Features: Occurs non-crystalline, filling cavities in rock and geodes; also in beds or "lakes." Color, black and brownish-black. Hardness = 1. Luster is pitchy. Generally melts in a match flame, burning and giving a bituminous odor.

Occurrence: In rock cavities and in the lining of geodes. In beds or "lakes" like the Trinidad deposit, Trinidad Island, off the coast of Venezuela.

Uses: Roofing compounds, street paving, preserving wood, etc.

AZURITE $Cu_3(OH)_2(CO_3)_2$
(Blue Carbonate of Copper)

Locality: Douglas County, in Ball mine on Cedar Springs Mountains. **Josephine County,** in oxidized ore. Queen of Bronze mine, Waldo district, and Almeda copper mine, Galice district. **Coos County,** upper Rock Creek.

Distinguishing Features: Azurite occurs in monoclinic crystals which are usually short prismatic or tabular in habit. Color, deep azure blue. Effervesces in cold HCL. Fuses on charcoal to metallic copper, yielding green flame.

Occurrence: In the oxidized zone occurring as a secondary mineral. Generally associated with malachite (green carbonate of copper) and at times found with limonite.

Uses: Ore of copper. Pigment (not very durable).

BALTIMORITE
(Var. of Serpentine)

Locality: Grant County, Canyon City.

NOTE.—Baltimorite is a variety of Serpentine having practically the same composition as Serpentine. It is a grayish-green mineral with a silky luster.

BARITE $Ba SO_4$
(Baryta, Heavy Spar, Baria)

Locality: Lane County, Bohemia mining district. **Josephine County,** Galice district, about 26 miles below Grants Pass. (Occurs as a gangue mineral in Almeda mine.)

Distinguishing Features: Barite occurs in crystals of the orthorhombic system, the tabular habit being most common. Three cleavages, two at right angles and one at oblique angle. Very heavy. Specific gravity = ±4.5. Generally white color, but often discolored by impurities. Insoluble in acids. Hardness = 2.5—3.0. Colors blowpipe flame yellowish-green.

Occurrence: Barite occurs as a gangue mineral in ore veins especially with lead ores. In cavities in limestone as a secondary mineral. Sometimes it occurs as lenticular masses in residual clays overlying limestone.

Uses: Manufacture of paint, being a substitute for white lead. Some varieties take a high polish and are used in place of marble. It is also used to give weight to paper.

BIOTITE $(H, K)_2 (Mg, Fe) Al_2Si_3O_{12}$
(Black Mica, Magnesium Mica)

Locality: **Jackson County,** in Gold Hill district on Upper Evans Creek, where it occurs in dike rocks.

Distinguishing Features: Occurs in platy and scaly masses, and sometimes in pseudohexagonal crystals. Cleavage is perfect basal, giving thin elastic sheets. The size of sheets ranges from small flakes to pieces 12 inches across. Color, black to dark brown. Thin sheets are translucent. Decomposed by concentrated H_2SO_4

Occurrence: In igneous rock, especially granites and a dike rock called lamprophyres. In mica-chist and gneiss.

Use: Same as muscovite.

BORAX $Na_2B_4O_7.10H_2O$
(Tinkal)

Locality: **Harney and Lake Counties,** marsh type of borax deposits have been worked. **Curry County,** five miles north of Chetco, where (pricite) a calcium borate mineral has been mined.

Distingushing Features: Borax occurs in monoclinic crystals, and in efflorescent crusts. Hardness = 2.—2.5. Specific gravity = ±1.7. Color, white or colorless. Fuses easily to a clear glass, giving intense yellow flame. Soluble in water. Has a disagreeable, alkaline taste.

Occurrence: In lake deposits in arid regions, where it occurs associated with gypsum, halite, hanksite, glauberite, thenardite, etc. San Bernardino County, California, in the Searles' borax marsh. In veinlets in sheared serpentine rocks.

Use: Large quantities are made use of in many ways in the arts and trades; i. e., as a flux in metallurgical processes, in manufacture of glass and gems, in soldering, and in making soap, toilet preparations, food preservatives, etc.

BORNITE $Cu_5 Fe S_4$
(Peacock Ore, Horse-Flesh Ore, Purple or Variegated Copper Ore)

Locality: Baker County, mined in Lower Snake River regions and in Copper Butte. **Josephine County,** less important ore in mines near Waldo. In Almeda mine, Galice district. Lower Applegate district. **Jackson County,** Ashland and Gold Hill districts. **Curry County,** Collier Creek district and McKinley Copper group.

Distinguishing Features: Bornite occurs in masses, rarely in cubical crystals. Color is dark reddish-brown on fresh fractures, soon tarnishing to purple. This feature will distinguish it from chalcocite. Fuses at 2½ on charcoal to a magnetic globule.

Occurrence: In veins associated with chalcopyrite and chalcocite. As a secondary enrichment product. As a contact mineral between igneous rocks and limestone.

Uses: Valuable ore of copper.

CALCITE $Ca CO_3$
(Calc-spar)

Locality: Curry County, veins in myrtle formation on Rogue River at Agness. **Jackson County,** Ashland, Upper Applegate, Jacksonville and Gold Hill districts. **Josephine County,** Galice, Almeda mine, Grants Pass, Lower Applegate districts and Oregon Caves. **Wasco County,** near Antelope. **Lane County,** on Farrington ranch, about two miles southwest of Eugene, where it occurs as nodules in basalt. **Baker County,** Sumpter. **Grant County,** 15 miles west of Mitchell, on South Fork of Taylor Creek, where it occurs in a nearly vertical vein, three feet thick and traced for 200 feet.

Distinguishing Features: Found in crystals of the rhombohedral and scalenohedral class of the hexagonal system; also occurs in crystalline crusts and druses and cleavable masses. Other forms are stalactitic, öolitic, pisolitic, granular and fibrous masses. Cleavage is perfect rhombohedral in three directions, at angles of 74° 55'. Parting is sometimes better developed than the cleavage. Polysynthetic twinning is sometimes developed, which appears as striations parallel to the long diagonal of the crystal. Hardness = 3. Specific gravity = ±2.72. Colorless when pure, but occurs in many colors due to impurities. Effervesces in cold acids. Distinguished from aragonite by cobalt nitrate test. (See Aragonite.)

Occurrence: In veins, often as a gangue mineral of ore bodies. In caves, as stalactites and stalagmites and as calcareous tufa and travertine. As the chief constituent of the shells and other hard parts of such organisms as molluscs, brachiopods, crinoids and corals. In cavities in basic igneous rocks. In seams of sedimentary rocks as a secondary mineral. Composes the greater part of crystalline limestones.

Uses: Limestone used in cement manufacture and as building stone. Road metal. Flux for smelters. Iceland spar is used in optical instruments; i. e., the polarizing microscope. A source of lime.

CELESTITE Sr SO₄

Locality: Josephine County, Almeda mine, Galice district.

Distinguishing Features: Occurs in crystals resembling barite in form and angle. It also occurs in fibrous seams and cleavable masses. Cleavage, perfect basal. Color, light blue, white and reddish. Hardness = 3. Specific gravity = ±3.9. Fuses fairly easily to a white pearl, giving a crimson red flame which is intensified by H Cl. Insoluble in acids. Dissolved in soda by fusion, which is in turn soluble in warm water, giving a white precipitate which is insoluble in H Cl.

Occurrence: In ore veins, as a gangue mineral. As a secondary mineral in limestone. Near Austin, Texas.

Uses: Manufacture of fire-works.

CERUSSITE Pb CO₃
(White Lead Ore)

Locality: Douglas and Lane Counties, in the Bohemia district.

Distinguishing Features: Occurs as silky prismatic masses; granular to compact massive; also as star shaped twins. Specific gravity = 6.5. Adamantine luster. Color, colorless, white, gray. Brittle. Fuses easily. On charcoal, fuses to a metallic lead button, depositing a yellow coating on the charcoal. Effervesces in hot HCL.

Occurrence: In oxidized ore veins as a secondary mineral derived from galena. Sometimes pseudomorphs after galena are found.

Uses: Ore of lead, carrying silver values.

CHALCEDONY Si O$_2$

Locality: Jackson County, near Eaglepoint. **Grant County,** John Day River near Dayville, John Day and Basin. **Wasco County,** near Antelope. **Harney County,** at many localities in the Harney desert. **Baker County,** Sumpter quadrangle. **Lane County,** Willamette and McKenzie River gravels. A very common mineral throughout the State.

Distinguishing Features: Occurs botryoidal, mammillary and stalactitic. More or less conchoidal fracture. Hardness $=7$. Color, colorless, white and many other colors due to impurities. Banded and variegated varieties are called agate. Red and brown varieties called jasper. Water agates are shells of chalcedony containing bubbles of water.

Occurrence: In seams and cavities in rocks, especially volcanic, igneous rocks occurring as a secondary mineral. Chert, flint and jasper varities occur in layers, lenses and concretions. in sedimentary rocks. Geodes are made up partially of chalcedony, which occurs in layers at the base of the in-pointing quartz crystals. Agates are very common on the Oregon beaches. The beach at Port Orford has furnished a large supply of beautiful stones. An agate carnival is held each year at Port Orford.

Uses: Jasper, agate and chrysoprase (apple green translucent chalcedony) are used for ornamental purposes. The color of chrysoprase is due to Ni O.

CHALCOCITE Cu$_2$S
(Copper Glance)

Locality: Douglas County, in Ball mine near Cedar Springs Mountains. **Baker County,** mined in Lower Snake River and Copper Butte regions. **Josephine County,** Waldo and Galice districts.

Distinguishing Features: Usually occurs in compact masses. Color, dark lead gray on fresh fracture, but tarnishes to dull black or green. Cuts easily when pure. No cleavage. Fuses easily on charcoal, giving a metallic button of copper. Does not become magnetic on heating; differs from bornite.

Occurrence: In ore veins with other sulphide minerals.

Uses: Copper ore.

CHALCOPYRITE Cu Fe S$_2$
(Copper Pyrites, Yellow Copper Ore, Fool's Gold)

Locality: Grant County, Quartzburg district, Susanville district and Copperopolis claims. Josephine County, Bradon and

Opp mines, Grants Pass district and Queen of Bronze mine. **Baker County,** Sumpter, Greenhorn, Elkhorn, Sparta and Cable districts. **Lane and Douglas Counties,** Bohemia and Blue River districts, Upper South Umpqua near Canyonville and Purdue. **Jackson County,** Gold Hill, Upper Applegate and Ashland districts. **Josephine County,** Galice and Waldo districts. **Wallowa County** near Lostine. **Curry County,** Mule Mountain district in gold veins, and Illinois River two miles north of mouth of Collier Creek. **Coos County,** Upper Rock Creek.

Distinguishing Features: Usually occurs massive. Sometimes in sphenoidal hemihedral tetragonal crystals. Color, bright brass-yellow. Tarnish, iridescent. Streak, greenish black. Brittle. Hardness = 3.5—4. (Pyrite = 6—6.5.) Fuses on charcoal to a brittle magnetic globule. Soluble in HNO_3 to a green solution with separation of sulphur. Sometimes occurs as a coating over tetrahedrite crystals.

Occurrence: In veins associated with galena, pyrite, sphalerite, tetrahedrite, siderite, cassiterite, bornite, chalcocite and pyrrhotite. Disseminated through massive pyrite. As a contact mineral with hematite or magnetite.

Uses: Ore of copper. Sometimes carries gold and silver values.

CHLORITE $H_3(Mg, Fe)_5 Al_2(SiO_6)_3$

Locality: Jackson County, Ashland district. **Josephine County,** Galice district.

Distinguishing Features: Occurs coarse to fine scaly, earthy and granular masses. Chlorite also occurs in disseminated scales, in monoclinic crystals of pseudohexagonal habit and as a pigment through other minerals and rocks. Cleavage is perfect basal. Cleavage lamellae are flexible, not elastic. (Distinction from mica.) Foliae are tough. Hardness = 2. Color, dark green to light green. Slight soapy feel to massive material.

Occurrence: Formed by alteration of silicate minerals like biotite, hornblende and augite in igneous rocks. As chlorite-schists, where it forms the major portion of the rock.

CHROMITE $Fe Cr_2 O_4$
(Chromic-iron Ore)

Locality: Grant County, small masses in serpentine two miles south of Cummings Ranch, John Day River, and seven miles south of Prairie City. **Douglas County,** Nickel Mountain near Riddles. **Baker County,** float in placers of Winterville, Upper

Burnt River and Bonanza districts. **Josephine County,** Waldo and Lower Applegate districts. **Curry County,** Sixes River, three-fourths mile above mouth of Dry Creek. **Coos County,** in elevated beach gravels between Three Mile Creek and The Lagoons. **Baker County,** Durkee, Sumpter (in black sand). **Clatsop County,** Astoria, Clatsop Beach, Hammond, near Seaside, Warrenton, Gearhart Beach, Fort Stevens, Carnahan Station, Morrison and Elk Creek (in black sand). **Coos County,** Marshfield, Bullards, South Fork Coquille River, Randolph district (old and present beaches), Whiskey Run, Whiskey River, Bandon Beach, Johnson Gulch and Johnson Mt. **Curry County,** Gold Beach, Chetco, Port Orford and Toledo beach, Rogue River beach, near Pistol River, beach sand at Ophir and Cuneffs Beach (in black sand). **Douglas County,** Glendale and Starvout (in black sand). **Grant County,** Granite, Vinson Creek and Big Creek (in black sand). **Jackson County,** Ashland, Gold Hill, Jacksonville, Medford, Birdseye Creek and Watkins (in black sand). **Josephine County,** Josephine Creek near Kerby, Holland, Kerby, Galice, Sucker Creek, Wolf Creek, Placer (in black sand). **Lincoln County,** Coos Bay (in black sand). Yaquina Bay, Newport and Toledo (in black sand). **Linn County,** Foster (in black sand). **Multnomah County,** Columbia River (in black sand). **Polk County,** Fall City (in black sand). **Umatilla County,** Weston (in black sand). **Wallowa County,** Wallowa (in black sand). **Wasco County,** Hood River beach at entrance to Columbia (in black sand). **Wheeler County,** Antone (in black sand).

Distinguishing Features: Occurs in masses, fine granular to compact. Color, black. Streak, dark brown. Fused with soda gives a megnetic mass. Borax and sodium metaphosphate beads are emerald green in both oxidizing and reducing flames of blowpipe. Insoluble in acids. Sometimes magnetic after heating.

Occurrence: In serpentine rocks due to alteration of the serpentine. As an original constituent in peridotites.

Uses: Chromite is the source of the chromium salts; i. e., potassium chromate, potassium dichromate and lead chromate. Also used in making hard steel and for making brick to line smelter furnaces in some instances.

CHRYSOCOLLA Cu Si O_3.$2H_2O$

Locality: **Josephine County,** is the oxidized ore in the Queen of Bronze mine, Waldo district. **Curry County,** Collier Creek Copper district. **Coos County,** head of Rock Creek.

Distinguishing Features: Occurs in smooth amorphous masses in seams and as incrustations. Often associated with other copper minerals. Color, bluish-green. Decomposed by HCL without effervescences or gelàtinization. (Malachite effervesces in HCL.) Fused with soda on charcoal, gives a copper button. In a closed tube, the mineral turns black and gives off water. On charcoal alone, it is infusible.

Occurrence: A secondary mineral associated with azurite, malachite and cuprite. Usually found in the upper workings of copper mines.

Uses: Ore of copper. (Oxidized ore.) Sometimes used as a poor substitute for turquois.

CHRYSOTILE $H_4 Mg_3 Si_2 O_9$
(Asbestos)

Locality: Josephine County, reported 10 miles west of Kerby and at Brownston. **Jackson County,** Buncomb, Kubli and Spikenard. **Lane County,** Meadow. **Grant County,** Mt. Vernon, Canyon City and Prairie City. **Lake County,** Lakeview. **Wheeler County,** Barite. **Malheur County,** Watson and Ontario. **Douglas County,** Perdue, Starveout, Canyonville and Crow Creek. **Clackamas County,** upper head waters of Clackamas River. **Jackson County,** in Gold Hill district. **Curry County,** in small seams scattered through the serpentine rocks.

Distinguishing Features: Fibrous; the fibers usually flexible and easily separating. Silky luster. Color, greenish-white, green, olive-green, yellow and brownish. It includes most of the silky amianthus of serpentine rocks.

NOTE.—There are two different minerals known in the trade as asbestos, one fibrous serpentine, or chrysotile, and the other fibrous amphibole. The chrysotile is the stronger of the two, but the fiber is shorter and coarser.

Occurrence: As a secondary mineral in veins and seams; also occurs on the borders of serpentine rocks and in veinlets in the serpentine.

Uses: Asbestos is used in making fireproof rope, cloth, boards and blocks. The asbestos is used alone or mixed with Portland cement, magnesia, saw dust, etc.

CINNABAR $Hg S$
(Mercury Blende)

Locality: Lane County, Black Butte. **Jackson County,** Palmer Creek near Ashland. **Grant County,** Granite and Susanville

districts. **Baker County,** Sumpter district. **Crook County,** How-
ard. **Douglas County,** formerly mined northeast of Oakland.
Jackson County, near Brownsboro and Meadows district; also
Ashland and Gold Hill districts. **Douglas County,** South Umpqua
River and Cow Creek. **Josephine County,** Picket Creek. **Curry
County,** Sixes River, in placers.

Distinguishing Features: Usually occurs massive and in
earthy forms through the rock. Kidney shaped masses or "Kid-
neys" occur in the Blackbutte mine, Lane County. Minute crystals in cavities are sometimes found. Color, scarlet to dark red.
Impure variety appears black. Streak, vermilion. Specific gravity = 8—8.2. Volatile. Gives a sublimate of metallic mercury
when heated in a closed tube with dry soda.

Occurrence: Occurs scattered through the rocks in stringers
instead of distinct veins. Kidney shaped masses in country rock
as seen in the Black Butte mines, Lane County. In the Ashland
district the cinnabar occurs in calcite veins in the bedrock of
placer mines.

Uses: The chief ore of mercury.

COBALTITE Co As S
(Cobalt Glance, Cobaltine)

Locality: **Grant County,** with gold and chalcopyrite in Quartzburg district, near Prairie City. **Josephine County,** float pieces
found near Kerby.

Distinguishing Features: Occurs generally granular to compact massive. Sometimes alters to erythrite, a pink arsenate of
cobalt. The erythrite forms a coating on the cobaltite. Color,
tin-white to steel-gray, sometimes with a pinkish tint. Fused on
charcoal, gives off arsenic fumes and sulphur. Hardness = 5½.
Specific gravity = ±6.1.

Occurrence: In veins and as float pieces. In schists and
gneisses. Tunaberg, Sweden, is an important foreign locality.

Uses: Ore of cobalt and used in porcelain painting.

COLEMANITE $Ca_2B_6O_{11} + 5H_2O$
(Priceite)

Locality: **Curry County,** Chetco (on coast five miles north of).

Distinguishing Features: Occurs usually as a crystal lining
of geodes. Also in compact crystalline masses. Color, white, colorless, yellowish, grayish. Perfect clino-pinacoidal (O1O) cleavage.
Easily fusible with exfoliation, coloring flame green. Soluble in
hot HCL.

Occurrence: In lake beds. Mineral at Chetco occurs in stringers along the shear zones in serpentine.

Uses: Source of borax and boracic acid.

COPPER (Ore)—Minerals

See Native copper, Bornite, Chalcocite, Chalcopyrite and chrysocolla.

COPPER (native) Cu

Locality: **Douglas County,** in Ball mine near Cedar Springs Mountain. **Curry County,** on Lower Illinois River and Collier Creek district. **Douglas County,** Dodson Butte district. **Baker County,** Sumpter quadrangle. **Josephine County,** Almeda mine, Galice district.

Distinguishing Features: Copper (native) is found in small disseminated grain, in sheets, and sometimes in large masses. Occasionally dendritic groups of distorted isometric crystals are found. Usually tarnished, but bright copper color can be seen by cutting with a knife. Copper is malleable. Under blowpipe fuses at (3) to metallic globule. Soluble in HNO_3, giving a green solution which becomes deep azure blue on the addition of NH_4 OH. Specific gravity = 8.8—8.9.

Occurrence: Occurs as a secondary mineral in mines where it has been formed by the reduction of copper compounds in solution. These copper compounds were in turn formed by the oxidation of chalcopyrite. Sometimes found in amygdaloidal cavities of diabase associated with calcite, zeolites, epidote, prehnite and datolite. This type is characteristic of the Upper Peninsula of Michigan. In ore veins copper (native), is generally associated with cuprite, azurite and malachite.

Uses: Ore of copper (the Upper Peninsula of Michigan being the one important locality).

CORUNDUM Al_2O_3

Locality: Josephine County, occurs in a small ledge near Grants Pass. Little work has been done on this deposit.

Distinguishing Features: Corundum occurs in loose crystals of the scalenohedral class, hexagonal system, the habit being pyramidal, tabular and prismatic. It also occurs in cleavable masses, also disseminated through rocks as small grains. Hardness = 9 (next to diamond). Specific gravity = ±4.0. Color, commonly bluish-gray. Other colors are brown, blue, red, pink and white. Infusible in blowpipe flame. When moistened with cobalt nitrate and heated, the mineral becomes deep blue. (Test for infusible alumina minerals; also zinc silicates.)

Occurrence: In syenites and nepheline syenites as disseminated crystals. Along the borders of peridotites and adjacent rocks. In crystalline limestones. In sands and gravels. (Gem-bearing gravels of Ceylon furnish sapphire and ruby.)

Uses: Gems, red variety is ruby, blue variety is sapphire. Used also as an abrasive either alone or as a mixture known as emery. Emery is made by mixing corundum, magnetite or hematite and sometimes spinel. Other gems, named according to color are: yellow, Oriental topaz; purple, Oriental amethyst; light to deep green, Oriental emerald.

CUPRITE Cu_2O
(Ruby Copper, Red Oxide of Copper)

Locality: **Josephine County,** oxidized ore of Queen of Bronze mine, Waldo district. **Douglas County,** in Ball mine, Cedar Springs Mountain.

Distinguishing Features: Crystals of the cubical, octahedral and dodecahedral forms are common. The crystals are usually small. Cuprite also occurs as crystalline aggregate and in fine grained masses. Specific gravity $= \pm6.0$. Color, dark red to a brownish red. Streak is brownish-red. Hardness $= 3\frac{1}{2}$—4. Harder than cinnabar and proustite, but softer than hematite. On charcoal fuses at ($2\frac{1}{2}$) to a copper button. Soluble in HNO_3, giving a green solution.

Occurrence: In the oxidized zone of ore bodies, where it occurs as a secondary mineral derived from copper compounds, especially native copper.

Uses: Copper ore.

DIATOMACEOUS EARTH
(Composed of siliceous remains of microscopical plants or animals called Diatoms)
(Infusorial Earth)

Locality: **Klamath County,** near Linkville. **Wasco County,** at Mosier. **Crook County,** Lowerbridge. **Grant County,** 20 miles north of Dayville on north bank of John Day River. **Baker County,** in townships 13 and 14 S., R. 37 E. **Union County,** four miles southeast of Elgin on Indian Creek.

Distinguishing Features: Earthy, non-effervescent in acids. Color, when pure, white. Cream colored material is common.

Best test for the material is to study it under the microscope. The diatom shells are easily recognized. A simple method for preparing material for study is to mount some of the powdered material in balsam on a glass slide.

Occurrence: In beds of various dimensions. The skeletons of diatoms have been replaced by silica.

Uses: Polishing powder. Fire-proof cement, boiler backing, mixed with clay and made into partition brick or tile. Mixed with cement giving what is known as "Puzzuolana," a substitute for Portland cement. Germany makes use of this material in the following ways: Artificial fertilizers, in manufacture of water glass, various cements, glazing for tile, artificial stone, sealing wax, fireworks, gutta-percha objects, Swedish matches, solidified bromine and papier-mache.

DOLOMITE Ca, Mg $(CO_3)_2$
(Magnesian Spar)

Locality: Baker County, Sumpter quadrangle.

Distinguishing Features: Dolomite occurs in rhombohedral hexagonal crystals closely resembling calcite, but having a lower grade of symmetry. It also occurs in granular and cleavable masses and in crystal druses. Occurs in warped rhombohedrons usually grouped with the sharp edges outward, giving the "saddle structure." Cleavage is rhombohedral, like calcite. Hardness = 3½—4. Specific gravity = 2.8. Color, pink, white or gray. Pearly or vitreous luster. Effervesces vigorously in warm acid. In cold acid the action is very slow (a distinction from calcite).

Occurrence: In cavities in limestone occurring as a secondary mineral. The chief constituent of dolomitic limestone. In veins associated with calcite.

Uses: Dolomitic limestones used for building stone, ornamental purposes and furnace lining. Dolomite, when burned, gives a lime which makes a very durable cement. Epsom salts are made from dolomite.

EPIDOTE HCa$_2$ (Al. Fe)$_3$ Si$_3$ O$_{13}$

Locality: Curry County, in placers of Sixes River. Coos County, in elevated beach gravels between Three Mile Creek and The Lagoons. **Douglas County, Roseburg quadrangle.**

Distinguishing Features: Occurs in monoclinic crystals often deeply striated. Crystals sometimes twin so as to appear orthorhombic. The mineral also occurs coarse to fine granular. Cleavage, perfect basal (001). Hardness $=6.-7$. Color, brown to greenish-black when in crystals. Green when massive.

Occurrence: In the contact zone between granite and limestone especially. Along seams of igneous rocks, especially granite. In schists as a metamorphis product.

ERIONITE $H_2Si_6Al_2$ Ca K_2 $Na_2O_{17}+5H_2O$
(New Zeolite)

Locality: Baker County, Durkee (in rhyolite tuff).

Distinguishing Features: Occurs in fine threads of a snow-white color. Luster, pearly. The threads are like soft wool, having a curly nature. Sometimes occurs as compactly matted fibers filling rock fissures. Occurs as white tufts adherent to the base of milky opal, and resembles the filamentous growth of opal. Some specimens have a thin coating of white opal. Mineral fuses easily and quietly in the flame to a clear glass. Water in relatively large amount is given off, when heated in a closed tube. This water has an alkaline reaction.

Reference: (See A. S. Eakle in reference list.)

FREIBERGITE
(See Tetrehedrite)

GALENA PbS
(Galenite, Lead Glance)

Locality: Grant County, Susanville, Quartzburg district and Alamo. Baker County, Sumpter, Baker, Greenhorn, Elkhorn, Cornucopia, Cable Cove and Sparta districts. Lane County, Blue River and Bohemia districts. Jackson County, Ashland, Upper Applegate and Gold Hill districts. Linn County, Santiam. Josephine County, Almeda mine, Galice district.

Distinguishing Features: Occurs in masses of cubical crystals; also in fine granular masses called steel galena. Cleavage is perfect cubical. Hardness $= 2\frac{1}{2}$. Specific gravity $= \pm 7.5$. Color, lead gray, often tarnished. Metallic luster. Fuses easily on charcoal to a lead button. A yellow sublimate (PbO) forms near the assay, and a white sublimate ($PbSO_4$) beyond the yellow one.

Occurrence: As a vein mineral associated with chalcopyrite, pyrite and sphalerite. In some localities barite, fluorite, and calcite are associates. Galena also occurs as a replacement of limestone due to metasomatic processes.

Uses: Ore of lead. At times contains enough silver to make it a valuable silver ore. Gold is often present.

GARNET (Ca, Mg, Fe, Mn)$_3$ (Al, Fe, Mn, Cr, Ti)$_2$ (SiO$_4$)$_3$

Locality: **Baker County,** Pleasant Valley; also in Sutter Creek district south of Baker. **Curry County,** Sixes River three-quarter mile above mouth of Dry Creek. **Coos County,** in elevated beach gravels between Three Mile Creek and The Lagoons. **Douglas County,** Roseburg quadrangle. **Baker County,** Durkee, Anthony and Richland (in black sand). **Clatsop County,** Astoria, Clatsop Beach, Hammond, Gearhart Beach, Warrenton, Fort Stevens, Carnahan Station, Seaside, Morrison, Clatsop Spit and Elk Creek (in black sand). **Coos County,** Marshfield, Bullards, South Fork, Coquille, Randolph district (old and present beaches) Whiskey Run and Johnson Gulch (in black sand). **Curry County,** Gold Beach, Port Orford, Rogue River Beach and Cuneffs Beach (in black sand). **Douglas County,** South Umpqua River (in black sand). **Grant County,** Vinson Creek and Big Creek (in black sand). **Jackson County,** Gold Hill (in black sand). **Josephine County,** Holland, Kerby, Waldo and Wolf Creek (in black sand). **Lincoln County,** Coos Bay, Yaquina Bay, Newport and Toledo (in black sand). **Linn County,** Foster (in black sand). **Malheur County,** Snake River (in black sand). **Multnomah County,** Latourell (in black sand). **Polk County,** Falls City (in black sand). **Tillamook County,** Oretown (in black sand). **Umatilla County,** Weston (in black sand). **Wallowa County,** Wallowa (in black sand). **Wasco County,** Hood River (in black sand).

Distinguishing Features: Garnet crystals are common. The crystals may be imbedded in a schist or in granular or in compact mineral masses. The crystals are usually dodecahedrons and trapezohedrons. Hardness = 7. Specific gravity varies from 3 to —4.3. Color, red, brown or yellow, rarely white, black or green. No apparent cleavage. Electrified when gently heated (i. e., will pick up small bits of paper). All varieties fuse at (4) except Uvarovite, which is infusable. There are several varieties of garnet among which are: Grossularite or Essonite Ca$_3$Al$_2$(SiO$_4$)$_3$, a white, yellowish, pale green or pale rose-red variety. Pyrope Mg$_3$Al$_2$ (SiO$_4$)$_3$, a deep red to nearly black variety. Almandite Fe$_3$ Al$_2$(SiO$_4$)$_3$, a deep red, transparent to brownish red or black variety. Spessartite Mn$_3$ Al$_2$(SiO$_4$)$_3$, dark purplish red or brownish red. Andradite, Ca$_3$Fe$_2$(SiO$_4$)$_3$, common garnet lacking transparency. Uvarovite Ca$_3$Cr$_2$(SiO$_4$)$_3$, emerald green, infusible variety.

Occurrence: In schists and gneisses (Almandite). In crystalline limestones at contacts with igneous masses. Associated minerals are wollstonite, diopside and vesuvianite (andradite and grossularite). In granites and granite-pegmatite (almandite and spessartite). In peridotites (pyrope). In nepheline and leucite-bearing lavas; for example, phonolites (andradite, variety melanite). In seams of chromite (uvarovite). In placers (commonly termed "Ruby Sands").

Uses: For abrasive purposes in finishing wood and leather. Some crystals are used for gems. Perfect crystals are valuable as museum specimens.

GARNIERITE H_2 (Ni, Mg), $SiO_4.H_2O$
(Genthite, Noumeite)

Locality: Douglas County, near Riddle. (Nickel Mountain three miles west of Riddle.)

Distinguishing Features: Occurs in earthy masses (in cellular chalcedony matrices at Riddle). Color, apple-green. Partially decomposed by HCL. Infusible. Becomes magnetic when heated on charcoal. Dull luster. Soft and friable. Resembles crysocolla, but the latter gives reactions for copper. Borax bead is violet when hot and reddish-brown when cold, after being heated in the oxidizing flame. Borax bead is turbid gray when heated in the reducing flame. NaOH added to nickel solutions gives a pale green precipitate insoluble in excess.

Occurrence: Occurs as a secondary mineral in chalcedony matrix associated with peridotite, Riddle, Oregon.

Uses: Ore of nickel.

GOLD (2) Au

Locality: Baker County, Baker, Canner Creek, Cornucopia, Durkee, Cracker Creek, Granite, Greenhorn, Mormon Basin, Pine Creek, Sumpter, Rye Valley, Unity, Weatherby and Whitney districts. **Curry County,** Mule Creek district. **Douglas County,** Bohemia, Canyonville, Cow Creek districts. **Grant County,** Quartzburg, Susanvile, canyonville, Granite, Alamo Canyon and Crane Creek districts. **Jackson County,** Applegate, Blue Ledge, Foots Creek, Galls Creek, Jacksonville, Draper, Ashland, Gold Hill and Pleasant Valley districts. **Josephine County,** Galice, Grave Creek, Louse Creek, Jump-Off Joe, Sucker Creek, Waldo, Davidson, Althouse and Wolf Creek districts. **Lane County,** Bohemia, Blue River and Fall Creek districts. **Malheur County,** Mormon Basin and Rye Valley. **Wheeler County,** Spanish Gulch district. **Marion County,** Santiam district near Detroit.

(2) Lode.

GOLD (1) Au

Locality: **Baker County,** Baker, Burnt River, Bridgeport, Cornucopia, Sanger, Canner Creek, Cracker Creek, Buck Creek, Durkee, Stice, Granite, Greenhorn, Deep Creek, Mormon Basin, Pine Creek, Rye Valley, Sumpter, Unity, Weatherby and Whitney districts. **Coos County,** Whiskey Run, Flanagan Bar, Johnson Creek, Myrtle Point, Eden and Randolph districts. **Crook County,** at Howard in black sands. **Curry County,** Ophir, Chetco, Gold Beach, Elk Creek, Mule Creek, Rogue River and Sixes River districts. **Douglas County,** Canyonville, Cow Creek, Green Mountain, Myrtle Creek, Ollala, Poker Flat and Perdue districts. **Grant County,** Beech Creek, Bull Run, Canyon Mountain, Elk Creek, Granite, Poker Flat, Susanville and Austen districts. **Jackson County,** Evans Creek, Foots Creek, Forest Creek, Jacksonville, Gold Hill, Draper, Woodville, Applegate, Rock Point and Weimer districts. **Josephine County,** Althouse, Sweat Basin, Wolf Creek, Applegate, Briggs Creek, Williams, Waldo, Sucker Creek, Rogue River, Picket Creek, Louse Creek, Jump-Off Joe, Illinois, Grave Creek, Galice and Winona districts. **Malheur County,** Mormon Basin district. **Wheeler County,** Spanish Gulch district. **Harney County,** Harney City. **Multnomah County,** in black sands at Portland and Latourell. **Baker County,** Anthony, Baker City, Huntington and Sparta (in black sand). **Clatsop County,** Astoria, Gearhart Beach, Seaside, Gearhart Park, Clatsop, Clatsop Spit and Elk Creek (in black sand). **Coos County,** Marshfield, South Fork Coquille River and Bandon Beach (in black sand). **Curry County,** near Pistol River, Eckley (in black sand). **Douglas County,** Drain, South Umpqua River, Steamboat River, Rogue River, Starvout, Glendale and Riddle (in black sand). **Grant County,** Homer, Vinson and Big Creeks (in black sand). **Jackson County,** Medford, Birdseye Creek and Watkins (in black sand). **Josephine County,** Holland, Kerby, Wolf Creek and Greenback, Myrtle Creek and Browntown (in black sand). **Lane County,** Cottage Grove (in black sand). **Lincoln County,** Yaquina Bay and Toledo (in black sand). **Linn County,** Foster (in black sand). **Marion County,** Detroit (in black sand). **Malheur County,** Snake River (in black sand). **Polk County,** Falls City (in black sand). **Umatilla County,** Weston (in black sand). **Union County,** La Grande (in black sand). **Wallowa County,** Wallowa (in black sand). **Wasco County,** Hood River (in black sand). **Washington County,** Hillsboro (in black sand). **Yamhill County,** North Yamhill (in black sand).

(1) Placer.

Distinguishing Features: Golden yellow color. Very malleable. Easily cut with a knife. Specific gravity = 15.6—19.3. Low specific gravity due to impurities. Easily fusible. Insoluble in acids. Soluble in aqua regia (a mixture of three parts concentrated HCL and one part concentrated HNO_3). Usually occurs finely disseminated through rocks; also in rounded grains and scales, and occasionally as nuggets. Octahedron crystals are not common. When found they are generally distorted, uniting to form a dendritic growth called wire-gold. With mercury forms amalgam.

Occurrence: In quartz veins associated with pyrite, chalcopyrite, sphalerite, arsenopyrite, etc. In placers. In quartz conglomerates as in the Rand, Transvaal, South Africa. As a secondary mineral in the oxidized zone.

Uses: The value and uses of gold are familiar.

GRAPHITE c
(Plumbago, Black Lead)

Locality: **Jackson County,** Upper Applegate district on hillside east of Sterling. Near Buncom and at the Blue Ledge mines in Upper Applegate district. **Josephine County,** Mayflower and Golden Wedge mines in Galice district.

Distinguishing Features: Occurs commonly foliated, granular, compact massive and as imbedded scales. Rarely forms tabular six-sided crystals. Cleavage, perfect basal. Color, dark gray to black. Feels greasy and rubs off easily on the fingers. Sectile. Cleavage laminae are flexible. Streak, dark gray. Streak on glazed porcelain, dark gray. (Molybdemite gives a greenish gray streak on glazed porcelain.) Specific gravity = 2.09—2.23. Infusible. Insoluble in acids.

Occurrence: In gneisses and schists. In fissure veins. In meteorites. In crystalline limestones.

Uses: Manufacture of lubricants, refractory crucibles, electric supplies and lead pencils (the hardness of the lead depending on admixed clay). Graphite can be made artificially in the electric furnace.

GYPSUM Ca $SO_4 + 2H_2O$

Locality: **Baker County,** mining and milling at Gypsum near Huntington. **Crook County,** near Bend, undeveloped deposit. **Wheeler County,** Bridge Creek. **Grant County,** John Day. **Crook County,** Crooked River. **Josephine County,** Almeda mine and Galice district. (Occurs as a gangue mineral.) **Jackson County,** Alton or Baron mine, Ashland district.

Distinguishing Features: Gypsum occurs in monoclinic crystals, in granular, earthy and fibrous masses, and in cleavable and crystalline masses. Luster is pearly, silky, dull or vitreous. Color, white when pure. May be gray, red, yellow, blue, brown or black due to impurities. Cleavage parallel to clino-pinocoid (010) perfect. Fair ortho-pinacoidal cleavage. Thin laminae are more or less flexible. Hardness = 1.5—2. Specific gravity = 2.32. Scratched with finger nail. Fuses easily at (3) to a white enamel giving a yellow flame. Soluble in HCL.

Occurrence: In lake bed deposits associated with salt, anhydrite and limestone. Gypsum is formed by the evaporation of the waters of inland seas. As a secondary mineral in various rocks formed by the action of H_2SO_4 upon calcium carbonates. As a hydration product of anhydrite.

Uses: Manufacture of plaster of Paris and as fertilizer. (Land Plaster.) As a retarder and adulterant in cement.

Varieties: Selenite.—Transparent crystals and cleavable plates or strips. Satin Spar.—Fibrous, silky-white material. Alabaster.—Fine granular material, white colored. Rock Gypsum.—Impure, dull colored, compact earthy to scaly material.

HALITE Na Cl
(Salt, Rock Salt, Common Salt)

Locality: Lake County, in salt marshes. **Douglas County,** Salt Springs south of Roseburg.

Distinguishing Features: Halite occurs in granular masses and in cubical hopper-shaped crystals. Color, colorless when pure. Various shades of yellow, brown and red, due to impurities. Cleavage, cubical. Taste is saline. Soluble in water.

Occurrence: In beds associated with anhydrite and gypsum. On the surface of dried salt lakes.

Uses: As table salt and as a preservative. Source of sodium compounds used chiefly in the manufacture of glass and soap.

Salt may be obtained: (1) By mining the rock salt as at Strassfort; (2) by pumping water down to the salt beds and then pumping the brine to the surface where the salt is recovered by the evaporation process; (3) by solar evaporation of seawater. The latter method is carried on at the Great Salt Lake, Utah; and at San Francisco, California; China and throughout the Orient.

HEMATITE Fe_2O_3
(Red Iron Ore)

Locality: **Lane County,** occurs in the ore mined for gold in Bohemia and Blue River districts. **Baker County,** specular variety is found in argillite, on Burnt River divide near Sumpter. **Jackson County,** in section 9, T. 38 S., R. 1 E., where it occurs in a clay bank; also in Gold Hill district. **Baker County,** Sparta, Sumpter quadrangle, Rye Valley and New Bridge (in black sand). **Crook County,** Howard (in black sand). **Curry County,** Eckley (in black sand). **Douglas County,** South Umpqua River, Steamboat River, Starvout and Riddle (in black sand). **Grant County,** Granite (in black sand). **Josephine County,** Josephine Creek near Kerby, Waldo, Sucker Creek and Browntown (in black sand). **Lane County,** Cottage Grove (in black sand). **Linn County,** Foster (in black sand). **Washington County,** Hillsboro (in black sand).

Distinguishing Features: Variety of forms—earthy, massive, micaceous, öolitic, fibrous and in hexagonal crystals. Streak is reddish-brown to red. Hardness $= 6$. Specific gravity $= \pm 5.2$. Difficult to fuse. Becomes magnetic on charcoal in reducing flame. Soluble slowly in concentrated HCL. Specular hematite is a micaceous variety made up of brilliant plates or crystals.

Occurrence: As a replacement of cherty iron carbonate by metasomatic processes. (Example is the Lake Superior hematite.) As a metasomatic replacement of öolitic limestone. (Clinton iron ore of Alabama is a fine example.) Formed from other minerals by alteration or replacement. Martite is a pseudomorphous form of hematite after magnetite (octahedral). Occurs as a pigment in menerals and rocks.

Uses: Ore of iron. For manufacturing of cheap paint, for chalk to mark assay crucibles and as a polishing powder.

HESSITE $(Au\ Ag)_2\ Te$

Locality: Baker County, in Sumpter district.

Distinguishing Features: Occurs compact and granular. Color, lead gray to steel gray. Streak, black. Specific gravity $= 8.31$—8.89. Fuses on charcoal to a black globule, dotted with silver. Powder this globule and drop the powder into boiling H_2SO_4, the acid will be colored purple.

Occurrence: As a vein mineral.

Uses: Rare but valuable ore of gold and silver.

HEULANDITE H_4 Ca Al_2 (Si O_3)$_6$+$3H_2O$

Locality: Lane County, Rowe River one-half mile down stream from Dorena.

Distinguishing Features: Occurs in monoclinic crystals. The crystals are usually thick tabular, parallel, to the side pinacoid (OlO). Crystals are often made up of many individuals, flattened parallel to the clino-pinacoid. Cleavage is perfect parallel to (OlO). The luster on these cleavage faces is pearly. Hardness = $3\frac{1}{2}$—4 . Specific gravity = ±2.1. Color, white to reddish. Fuses easily bubbling and swelling into a white mass. Gives off considerable water when heated in a closed tube. Slowly decomposed by HCL without gelatinizing.

Occurrence: In cavities in basic igneous rock, where it occurs as a secondary mineral associated with zeolites, such as stilbite, chabazite, etc. It has also been found in gneisses and in metalliferous veins.

ILMENITE (Fe Ti)$_2$ O$_3$
(Titanic Iron Ore, Menaccanite)

Locality: Curry County, in placer gravels of Sixes River. Coos County, in elevated beach gravels between Three Mile Creek and The Lagoons. Baker County, Durkee, Baker City, Sumpter, Sparta, Richland (in black sand). Benton County, Alsea (in black sand). Clatsop County, Hammond, near Seaside and Warrenton (in black sand). Coos County, Marshfield, Bullards, Whiskey River and Bandon beach (in black sand). Lincoln County, Coos Bay, Yaquina Bay, Newport and Toledo (in black sand). Crook County, Howard (in black sand). Douglas County, Drain, South Umpqua, Steamboat River, Rogue River, Glendale and Riddle (in black sand). Grant County, Vinson Creek and Big Creek (in black sand). Jackson County, Ashland, Weimer, Gold Hill, Foote Creek and Watkins (in black sand). Josephine County, Josephine Creek, Galice, Waldo, Sucker Creek, Wolf Creek, Placer and Kerby district (in black sand). Linn County, Foster (in black sand). Malheur County, in Snake River (in black sand). Marion County, in southeast corner (in black sand). Multnomah County, Fulton, Latourell, Columbia River sand, Portland (in black sand). Tillamook County, Oretown (in black sand). Union County, La Grande (in black sand). Wheeler County, Antone (in black sand). Yamhill County, North Yamhill (in black sand).

Distinguishing Features: Occurs granular and compact massive; also in thin plates and as sand. Hardness = 5—6.

Specific gravity = ±4.7. Color black. Streak, reddish-brown to black. Infusible. Very finely ground ilmenite fused with soda on charcoal and boiled in HCl till dissolved, will give a violet colored solution if tin is added and the solution boiled a little longer. (This is a test for titanium.)

Occurrence: In igneous rocks, chiefly diabases. In black sands.

Uses: Is used in lining pudding furnaces.

NOTE.—Due to the difficulty in smelting titanium-bearing iron ores, large bodies of such ore remain undeveloped in the United States and Canada.

INFUSORIAL EARTH
(See Diatomaceous Earth)

IRIDOSMINE Ir Os sometimes with Rh, Pt, etc.

Locality: Curry County, on Sixes River three-quarter mile above mouth of Dry Creek. Coos County, in elevated beach gravels between Three Mile Creek and The Lagoons. Josephine County, Lower Illinois River.

Distinguishing Features: Occurs commonly in irregular flattened grains. Rarely in hexagonal prisms. Hardness = 6—7. Specific gravity = 19.3—21.12. Metallic luster. Color, tin white and light steel gray. Difficultly malleable. Infusible. Insoluble.

Occurrence: Alluvial deposits, with gold, platinum, chromite, corundum, zircon and diamond.

IRON Fe
(See Limonite, Hematite, Magnetite and Siderite)

JOSEPHINITE $Fe_2 Ni_5$

Locality: Josephine County, in placer of Josephine Creek.

Distinguishing Features: Is a nickel-iron mineral from Josephine Creek, Oregon, occurring in the stream gravel. This mineral comes under the head of Terrestrial Iron. Silver color and rather heavy.

Occurrence: In stream gravels.

Uses: Valued as a mineral specimen.

KAOLINITE $H_4 Al_2 Si_2 O_9$
(Kaolin, China-Clay)

Locality: Malheur County, 28 miles southwest of Huntington. Coos County, Arago. Wasco County, five miles above The Dalles. Harney County, south of Lake Harney. Jackson County, 10 miles east of Ashland.

Distinguishing Features: Occurs usually in compact clay-like masses. Hardness = 2—2½. Specific gravity = 2.6. Color, white when pure, gray and yellowish due to impurities. Dull luster. Infusible if pure. Heated on charcoal after moistening with cobalt nitrate, the mineral becomes deep blue. Gives water in a closed tube. Insoluble in acids. Strong earthy odor when breathed upon. Often smooth to the touch.

Occurrence: As a secondary mineral due to alteration of aluminous silicates.

Uses: Manufacture of china, porcelain, stoneware, fire-brick, fancy tile, etc. (Kaolin is a mixture of kaolinite and aluminum silicates with more or less quartz, feldspar, etc.)

LEAD
(See Galena, Cerussite)

LIMONITE $Fe_2(OH)_6 Fe_2O_3$
(Brown Iron Ore)

Locality: Baker County, Baker City (in black sand). In oxidized zone in lodes in Sumpter quadrangle. Clackamas County, Oswego. Columbia County, Scappose. Douglas County, Steamboat River (in black sand). Jackson County, in Section 3, T. 36 S., R. 3 W., in Gold Hill district. Wheeler County, Antone (in black sand). Curry County, Wake-Up-Riley Ridge.

Distinguishing Features: Limonite occurs in earthy masses, in mammillary, botryoidal and stalactitic forms; also in fibrous, compact, pisolitic and nodular masses. Occurs as pseudomorphs after pyrite, especially the cubical form. Color, yellow, brown or black. Streak, yellow and yellowish brown. Magnetic on fusing in blowpipe flame.

Occurrence: In marshy places or bogs (bog-iron ore) and having a loose or porous texture. As a secondary mineral formed by the alteration of pyrite in ore veins. Comprises a large part of the gossan or "Iron-hat." As a stain in rocks and clays.

Uses: Ore of iron. Manufacture of some yellow and brown paints.

MAGNETITE Fe_3O_4

Locality: Jackson County, Gold Hill (occurs as a magnetite pyroxenite). Josephine County Waldo district (occurs in serpentine rock). In the Lower Applegate district, of above county, both magnetite and hematite occur as a cement in quartzite. Curry County, Gold Beach (in black sand); also on Sixes River three-quarter mile above the mouth of Dry Creek and one-half mile northeast of Horse Sign Butte. **Coos County,** in elevated beach gravels between Three Mile Creek and The Lagoons. **Baker County,** Durkee, Anthony, Baker City, Sumpter, Sparta, Rye Valley, Richland and New Bridge (in black sand). **Benton County,** Alsea (in black sand). **Clatsop County,** Astoria, Clatsop Beach, Hammond, near Seaside, Warrenton, Gearhart Beach, Fort Stevens, Carnahan Station, Clatsop Spit, Columbia River and Elk Creek (in black sand). **Coos County,** Marshfield, Bullards, South Fork Coquille River, Randolph district (old and present beach sands), Whiskey Run, Whiskey River, Bandon Beach and Johnson Gulch (in black sand). **Lincoln County,** Coos Bay, Yaquina Bay, Newport and Toledo (in black sand). **Crook County,** Howard (in black sand). **Curry County,** Gold Beach, Chetco, Ophir, Port Orford, Rogue River beach, near Pistol River, Eckley and Cuneffs Beach (in black sand). **Douglas County,** Drain, South Umpqua River, Steamboat River, Rogue River, Glendale, Starvout, Riddle (in black sand). **Grant County,** Comer, Granite, Vinson Creek and Big Creek (in black sand). **Jackson County,** Ashland, Weimer, Gold Hill, Jacksonville, Medford, Birdseye Creek, Foote Creek and Watkins (in black sand). **Josephine County,** Josephine Creek near Kerby, Holland, Kerby, Galice, Sutler Creek, Waldo, Sucker Creek, Wolf Creek, Placer, Coyote Creek, Greenback, Illinois River near Kerby, Browntown (in black sand). **Lane County,** Cottage Grove (in black sand). **Linn County,** Foster (in black sand). **Malheur County,** Snake River (in black sand). **Marion County,** in southeast part of county (in black sand). **Multnomah County,** Fulton, Latourell, Columbia River sand (in black sand). **Polk County,** Falls City (in black sand). **Tillamook County,** Oretown (in black sand). **Umatilla County,** Weston (in black sand). **Union County,** La Grande (in black sand). **Wallowa County,** Wallowa (in black sand). **Wasco County,** Hood River beach (in black sand). **Washington County,** Hillsboro (in black sand). **Wheeler County,** Antone (in black sand). **Yamhill County,** North Yamhill (in black sand).

Distinguishing Features: Occurs generally in black sand in placer deposits; also occurs in loose crystals and granular and compact masses. The octahedron, dodecahedron and trapezohedron are the common crystal forms. Hardness = 6. Specific gravity = ±5.1. Color, black. Metallic luster. Strongly attracted by magnet. Fuses with difficulty. Soluble in concentrated HCl. Streak, black.

Occurrence: In black sands of the Pacific Coast. Widely distributed through igneous rocks. In ore-deposits formed by magmatic segregation. In the contact zone between igneous rocks and limestone. In schists and gneisses as lens-shaped bodies.

Uses: Ore of iron.

MALACHITE $Cu_2(OH)_2CO_3$
(Green Carbonate of Copper)

Locality: **Baker County,** Copper Queen mine, Copper Butte district; also Sumpter quadrangle. **Grant County,** Copperopolis claims, Quartzburg district. **Douglas County,** Ball mine. **Josephine County,** Queen of Bronze mine, Waldo district, and Almeda mine, Galice district. **Jackson County,** Alton or Barron mines, Ashland district. **Coos County,** Upper Rock Creek.

Distinguishing Features: The characteristic occurrence of malachite is in mammillary crusts, fibrous masses and acicular monoclinic crystals. Color, bright emerald-green. No cleavage. On charcoal fuses easily to metallic copper, giving a copper (green) flame made azure blue by HCl. Effesvesces in cold HCl. (Distinction from chrysocolla, atacamite, and olivenite.)

Occurrence: In the upper oxidized zone of copper deposits as a secondary mineral. Constitutes the "copper stain." Malachite and azurite are common associates.

Uses: Ore of copper. Banded varieties are polished for ornamental purposes and jewelry. Pigment in imitation bronze.

MANGANESE
(See Pyrolusite, Rhodonite, Rhodrochrosite)

MARCASITE $Fe S_2$
(White Iron, White Pyrites, Spear Pyrites)

Locality: **Baker County,** Sumpter district. **Grant County,** in gold-quartz veins in Quartzburg and Granite districts. **Jackson County,** Ashland district.

Distinguishing Features: Marcasite occurs in orthorhombic crystals usually tabular in habit and elongated in the direction of the (a) axis; also in crystalline aggregates and rounded concretionary masses. Color, pale brass yellow with a greenish tinge. Metallic luster. On charcoal burns with a blue flame, giving off SO_2 and leaving a magnetic residue. Specific gravity $= 4.87$ (lower than pyrite). The crystal form distinguishes it from pyrite. Marcasite is more liable to decompose than pyrite.

Occurrence: In sedimentary rocks, often concretionary in form. In ore veins, but not as common as pyrite.

Uses: Could be used in the manufacture of sulphuric acid if a large enough deposit were available.

MICA
(See Muscovite and Biotite)

MIRABILITE $Na_2 SO_4 + 10H_2O$
(Glauber Salt)

Locality: Lake County, in ponds and lakes.

Distinguishing Features: Forms in crusts and as a powder. Specific gravity $= \pm 1.5$. Hardness $= 1\frac{1}{2}$—2. Color is white. Fuses easily, giving an intense yellow flame. Gives a great deal of water in a closed tube. Soluble in water.

Occurrence: In soda lakes. Forms along the shores of Great Salt Lake during the winter. Also occurs as an efflorescence in caves.

MOLYBDENITE $Mo S_2$

Locality: Baker County, Copper Creek in the Eagle Mountains, and Deer Creek near Sumpter. Josephine County, Blue Bell mine, Galice district. Jackson County, Ashland district south of Ashland. Wallowa County, Wallowa district near Aneroid Lake.

Distinguishing Features: Occurs generally in foliated masses and disseminated scales. Sometimes in hexagonal tabular crystals. Cleavage parallel to base. Laminae very flexible, but not elastic. Color, bluish lead gray. Streak on glazed porcelain or glazed paper has a greenish tinge. Infusible. Na PO_3 bead is green in reducing flame and colorless in oxidizing flame. Soluble in HNO_3, giving a white sublimate which is soluble in NH_4OH. Specific gravity $= 4.7$—4.8. Graphite is 2.09—2.23.

Occurrence: In granites and pegmatites. In veins with topaz, cassiterite and wolframite. In contact between limestone and granites associated with chalcopyrite, epidote, etc.

Uses: Ore of molybdenum, which is used as an alloy with steel. Makes a harder and tougher steel. The source of molybdenum salts used in analytical work.

MONAZITE (Ce, La, Di) PO_4

Locality: Clatsop County, Astoria, Clatsop Beach, Gearhart Beach, Hammond, Fort Stevens, Warrenton, Seaside, Gearhart

Park, Clatsop, Carnahan Station, Morrison, Clatsop Spit and Elk Creek (in black sand). **Coos County,** South Fork Coquille River, Randolph district (in black sand). **Curry County,** Gold Beach, Port Orford and Eckley (in black sand). **Josephine County,** Holland, Sucker Creek, Wolf Creek and Placer (in black sand). **Linn County,** Foster (in black sand). **Multnomah County,** Fulton and Latourell (in black sand). **Polk County,** Falls City (in black sand). **Umatilla County,** Weston (in black sand). **Wallowa County,** Wallowa (in black sand). **Wasco County,** Hood River (in black sand). **Wheeler County,** Antone (in black sand).

Distinguishing Features: Usually found in the form of sand. Monoclinic crystals are very small and not common. Hardness = 5. Specific gravity = ±5.1. Color, yellow, brown or brownish red. Streak is white. Very good basal cleavage. Infusible, but turns gray on heating. Decomposed by acids.

Occurrence: In granites and gneisses as an accessory mineral. In sands.

Uses: Monazite is the source of thoria (ThO_2) from which the Welsbach gas mantle is made. It is also the source of other rare earths. Source of Helium gas, extracted by means of great temperature and pressure in laboratory of University of Leiden, Holland.

MUSCOVITE H_2 (K Na) Al_3 (Si O_4)$_3$
(Isinglass, Potash Mica)

Locality: Jackson County, in Gold Hill district on upper Evans Creek, where it occurs in pegmatite dikes. **Douglas County,** Roseburg quadrangle.

Distinguishing Features: Occurs in cleavable and scaly masses. Rarely in pseudo hexagonal crystals. Cleavage is perfect basal (OOl), giving very thin sheets, which are elastic. For size of sheets see Biotite. Color, colorless to pale brown. Thin sheets are always transparent. Decomposed by H_2SO_4.

Occurrence: In granite-aplites and pegmatites. In mica-schist and gneiss. In sandstones. As a secondary mineral sericite derived from feldspar.

Uses: For electrical insulator. For stove door windows. For spangling paper and fabrics. Covering for steam boilers and pipes. For decorative interior work. To ornament porcelain and glassware. In calico printing. As a lubricant. As an absorbent of nitro glycerine. In manufacture of smokeless powder.

NATRON $Na_2 CO_3 + 10H_2O$

Locality: Lake County, in lakes and ponds.

Distinguishing Features: Occurs in powdery crusts or earthy masses. Effervesces in cold HCl. Specific gravity $= \pm 1.44$. Hardness $= 1—1.5$. Very brittle. Alkaline taste. Occurs in nature in solution or mixed with other carbonates of soda.

Occurrence: Lake deposits.

NICKEL
(See Garnierite)

SODA NITRE $Na NO_3$
(Chile Saltpeter, Cubic Niter)

Locality: Lane County, ledge near Mount June. In marsh deposits in southeastern Oregon.

Distinguishing Features: Occurs in granular masses in thick beds; also as efflorescent crusts associated with gypsum, halite, and other soluble salts. When exposed to the atmosphere, crumbles. Hardness $= 1\frac{1}{2}—2$. Color is white, gray or yellow. Tastes salty and cooling. Under blowpipe flame the mineral deflagrates and changes to a liquid. Dissolves easily in water.

Occurrence: In beds associated with halite, gypsum, etc. In ledges.

Uses: Manufacture of niter, nitric acid and fertilizers. When containing sodium iodide it becomes a source of commercial iodine. Manufacture of gun powder.

OLIVINE $(Mg Fe)_2 Si O_4$
(Chrysolite, Job's Tears)

Locality: Baker County, Durkee (in black sand). Clatsop County, Astoria, Gearhart Beach, Warrenton, Hammond, Fort Stevens, Carnahan Station, Seaside, Clatsop, Morrison, Clatsop Spit, and Elk Creek (in black sand). Coos County, Randolph district in old and present beach sand, Whiskey Run and Johnson Gulch (in black sand). Curry County, Gold Beach, Chetco, Port Orford, Rogue River beach and Eckley (in black sand). Josephine County, Holland, Kerby and Wolf Creek (in black sand). Multnomah County, Fulton and Latourell (in black sand. Polk County, Falls City (in black sand). Umatilla County, Weston (in black sand). Wasco County, Hood River beach (in black sand).

Distinguishing Features: Occurs in granular masses or disseminated crystals and grains. Crystals are generally tabular in habit and belong to the orthorhombic system. Hardness $= 6\frac{1}{2}$— 7. Specific gravity $= 3.3$. Color, yellowish green to light green. Infusible. Gelatinizes with H Cl. Luster, resinous (oily).

Occurrence: In peridotites associated with enstatite or diallage. In many cases the olivine is partially altered to serpentine. Dunite is an igneous rock composed almost entirely of olivine. In gabbro, basalt and diabase. In tuffs. In meteorites. In river and beach sands.

Uses: Clear variety, called peridot, is used as a gem.

OPAL $Si\ O_2 + nH_2O$

Locality: **Baker County,** near Durkee. **Gilliam County,** Hay Creek. **Grant County,** Grub Creek, near Canyon City. **Wasco County,** The Dalles and near Antelope. **Umatilla County,** four miles above Weston on Pine Creek. **Clackamas County,** Hyalite found in seams in trap rock forming falls at Oregon City.

Distinguishing Features: Occurs in cavities and seams, sometimes disseminated and massive. It is amorphous silica. Conchoidal fracture. No cleavage. Brittle. Hardness $= 5\frac{1}{2}$—$6\frac{1}{2}$. Specific gravity $= \pm 2.1$. Insoluble in ordinary acids. Soluble in HF or KOH. Color, white or almost any color. Greasy luster. Different varieties of opal are: Precious opal—milky, blue, yellow, red or black, showing internal reflections of various colors. Fire opal—shows reflected colors but not a play of colors like precious opal. Fire opal is red, yellow or brown. Wood opal— wood replaced by opal. May show the woody structure. Opal-Agate—opal with a structure like agate. Hyalite—a colorless, transparent variety occurring in a botryoidal coating or drops.

Occurrence: In cavities and seams in igneous rocks. Trachyte is the home of fire opal. Formed in hot spring deposits. Another mode of occurrence is in diatomaceous earth in which the diatoms secrete opal silica in their casts.

Uses: Gems (Precious and fire opal).

PLATINUM Pt

Locality: **Coos County,** Bandon and Whiskey River (in black sand). **Curry County,** Cape Blanco and Ophir. **Baker County,** near Durkee. **Douglas County,** Riddle, Drain and Glendale. **Josephine County,** Sucker Creek, Illinois River, Kerby district, Browntown and Grants Pass (in black sand), Galice, Waldo and Greenback districts. **Linn County,** Foster. **Union County,** La

Grande. **Wheeler County,** Antone (Spanish Gulch) placers. **Coos County,** Randolph district. **Curry County,** Upper Sixes River, Gold Beach and in placers of Rogue River.

Distinguishing Features: Occurs as scales, grains and nuggets in alluvial deposits. Rarely in isometric cubical crystals. Color, light steel-gray. Hardness = 4—4.5. Malleable and ductile. Specific gravity = ±14—19; 22 when chemically pure. Infusible. Soluble in hot aqua-regia only.

Occurrence: In placers with gold, magnetite, ilmenite, zircon. diamond and other heavy minerals. In peridotites. The platinum of the Urals and British Columbia are derived from this class of rock. In the Waldo district, Josephine County, the platinum usually occurs alloyed with a little iridium and osmium.

Uses: The platinum of commerce is derived from native platinum. This metal is used extensively in making laboratory apparatus and jewelry.

PROUSTITE Ag₃As S₃
(Ruby or Red Silver)

Locality: Baker County, Sumpter quadrangle.

Distinguishing Features: Occurs in compact masses; also as dissemiated grain through gangue and as a crust or stain. Rarely in small hexagonal crystals. Hardness = 2—2.5. Specific gravity = 5.5—5.6. Luster, adamantine. Streak, scarlet. Color, scarlet. Brittle. Fuses on charcoal giving off garlic and sulphurous odors and forming a silver bead. Decomposed by H NO₃. The powdered mineral is turned black by potassium hydroxide. Differs from pyrargyrite by having a scarlet streak. Cuprite and cinnabar do not give off the garlic odor when heated.

Occurrence: In veins with other minerals such as pyrargyrite, native silver, native gold and cerargyrite.

Use: Valuable ore of silver.

PYRARGYRITE Ag₃ Sb S₃
(Ruby Silver)

Locality: Grant County, Granite district. **Baker County,** Sumpter, Rye Valley and Elkhorn districts.

Distinguishing Features: Occurs massive and as hexagonal prismatic crystals. Hardness = 2½. Specific gravity = ±5.8. Color, black, but red by transmitted light. Streak, purplish gray. Fuses easily on charcoal to a globule of silver sulphide, forming a white sublimate. This globule can be reduced to silver by fusing with soda in the reducing flame. Decomposed by HNO_3 with the formation of sulphur and a white residue.

Occurrence: Vein mineral.

Uses: Ore of silver.

PYRITE Fe S$_2$
(Iron Pyrites, Fool's Gold)

Locality: Found in nearly all the mining districts of **Grant, Baker** and **Union Counties. Jackson County,** Bradon, Upper Applegate, Opp, Ashland, Tin Pan, Gold Hill and Jacksonville districts. **Josephine County,** Queen of Bronze mine, near Waldo; also Almeda mine, Galice district; Grants Pass and Lower Applegate districts. **Lane County,** Blue River district. **Douglas** and **Lane Counties,** Bohemia district. **Douglas County,** Roseburg quadrangle. **Curry County,** in gold veins of Mule Mountain district, Wake-Up-Riley Ridge and Illinois River two miles north of mouth of Collier Creek.

Distinguishing Features: Often found in isometric crystals of cubical, pyritohedral, diploidal and octahedral class. Twin crystals of penetration pyritohedron and cube are sometimes found. Hardness = 6—6.5. Specific gravity = ±4.95—5-10. Color, brass yellow. Tarnishes to brownish. No apparent cleavage. Fuses at (3) on charcoal to a magnetic globule, giving off SO_2. Insoluble in HCL, but soluble in HNO_3 with separation of sulphur.

Occurrence: As a vein mineral. As a secondary mineral in igneous rocks bordering ore deposits. Disseminated through limestone and shale. In metamorphic rocks as bedded deposits. As a contact mineral in the ores at contacts between sedimentary and igneous rocks.

Uses: Manufacture of sulphuric acid. Gold-bearing pyrite forms gold ore; at times also a low grade ore of copper when copper bearing.

PYROLUSITE Mn O$_2$
(Gray or Black Oxide of Manganese)

Locality: Baker County, Sumpter district. **Josephine County,** on Cave Creek in Waldo district, where the pyrolusite occurs as a surface alteration of rhodonite and rhodochrosite. **Jackson County,** Ashland district, Gold Hill district.

Distinguishing Features. Occurs in acicular crystals which are indistinct and probably pseudomorphous after manganite. Also occurs in crusts, in masses, in fibrous and columnar forms.

Very soft (1-2). Specific gravity = ±4.8. Metallic luster. Color, black to dark steel-gray. Streak, black, often sooty. Brittle. Infusible. Gives bead test for manganese. (Borax or Na PO_3 bead is amethyst in the oxidizing flame and colorless in the reducing flame.) Soluble in HCL, giving off chlorine.

Occurrence: In residual clays formed by the breaking down of limestone, giving a concentration of the manganese oxide. Along seams in rocks occurring as a secondary mineral in gold veins. As an alteration product of rhodonite and rhodochrosite.

Uses: Ore of manganese. As a decolorizer in glass manufacture. Colors glass purple. Manufacture of electric dry-cells. Manufacture of chlorine and oxygen. As an oxidizing agent.

PYROXENE Ca Mg Al Fe $(Si O_3)_2$
(Var. Augite)

Locality: Lane County, from croppings of dike one-half to three-fourths mile south of point on railroad track about three-fourths mile south of Goshen, Oregon.

Distinguishing Features: Occurs generally in monoclinic, prismatic crystals with a nearly octagonal or square cross-section. Color, light green to dark green; also white or brown. Cleavage, prismatic at angle of 87° 10'. Fusibility varies from easy to difficult, sometimes giving a magnetic globule. Hardness = 5—6. Specific gravity = 3.2—3.6.

Occurrence: In crystalline limestone, contact rocks, igneous rocks, crevices in hornblende or chlorite schist, serpentine, etc. Associated with chlorite, garnet, magnetite, pyrite, calcite, amphibole and wollastonite. In meteorite.

PYRRHOTITE $Fe_n S_{n+1}$
(Magnetic Pyrites)

Locality: Baker County, Virtue and Sumpter districts. Jackson County, Bradon, Jewet and Corporal G mines. Josephine County, Queen of Bronze mine, near Waldo. Jackson County, Ashland, Upper Applegate and Gold Hill districts. Josephine County, Galice and Grants Pass districts. Douglas County, Cow Creek Canyon.

Distinguishing Features: Occurs usually massive and without cleavage. Pseudohexagonal orthorhombic crystals are rare. Hardness = 4. Color, bronze-yellow. Tarnishes. Magnetic before heating. (This property distinguishes it from minerals it may resemble.) Soluble in HNO_3.

Occurrence: In basic igneous rocks. Many of the large deposits are due to magmatic segregation. As a vein mineral. In crystalline limestones.

Uses: Nickel-bearing pyrrhotite is an ore of nickel.

QUARTZ SiO_2

Locality: Jackson County, Ashland, Upper Applegate, Jacksonville, Gold Hill districts. Occurs as a gangue mineral. Josephine County, Galice, Grants Pass, Lower Applegate, Waldo districts. Occurs as a gangue mineral. Wasco County, near Antelope. Grant County, Canyon City district and Granite. Baker County, Sumpter. Douglas County, Roseburg quadrangle and Bohemia mining district. Lane County, Bohemia, Blue River and Fall Creek mines. Baker County, Durkee, Anthony, Huntington, Sparta, Rye Valley (in black sand). Benton County, Alsea (in black sand). Clatsop County, Astoria, Warrenton, near Seaside, Gearhart Beach, Hammond, Fort Stevens, Carnahan Station, Clatsop, Morrison, Clatsop Spit and Elk Creek (in black sand). Coos County, Marshfield, Bullards, South Fork Coquille River, Randolph district, Whiskey Run and Johnson Gulch (in black sand). Curry County, Gold Beach, Chetco, Port Orford, Rogue River beach, near Pistol River and Ophir (in black sand). Douglas County, Steamboat River, Glendale and Starvout (in black sand). Grant County, Vinson and Big Creeks (in black sand). Jackson County, Medford, Foote Creek and Watson (in black sand). Josephine County, Josephine Creek near Kerby, Holland, Kerby, Sutler Creek, Sucker Creek, Wolf Creek and Placer (in black sand). Lane County, Cottage Grove (in black sand). Lincoln County, Coos Bay, Yaquina Bay, Newport (in black sand). Linn County, Foster (in black sand). Malheur County, Snake River (in black sand). Marion County, southeast corner (in black sand). Multnomah County, Fulton and Portland (in black sand). Polk County, Falls City (in black sand). Umatilla County, Weston (in black sand). Wallowa County, Wallowa (in black sand). Wasco County, Hood River (in black sand). Washington County, Hillsboro, (in black sand). Wheeler County, Antone (in black sand).

Distinguishing Features: Occurs in hexagonal crystals made up of prisms terminated by pyramids. Also occurs massive and as sand. No cleavage. Hardness = 7. Specific gravity = ±2.66. Color, colorless when pure. May be shades of red, yellow, rosecolor, black and brown. Conchoidal fracture. Infusable. Insoluble in ordinary acids. Soluble in hydrofluoric acid (HF).

Occurrence: In granites, syenites and other igneous rocks such as quartz-basalts, rhyolites, etc. As a gangue mineral in ore veins. In sandstones and quartzites. In river and beach sands. As a secondary mineral in various rocks. As a lining of geodes.

Uses: Rock crystal for optical instruments. Rock crystal, amethyst and smoky quartz for ornamental purposes. Quartz sand for glass making. With coke to make carborundum. As sandstone for hones and grinding stones. Fused in electric furnace for making heat-resisting "silica ware." As a flux in metallurgy.

QUICKSILVER
(See Cinnabar)

RHODONITE Mn Si O₃ with replacement by Fe, Zn or Ca
(Manganese Spar)

Locality: Josephine County, Waldo district on Cave Creek, three miles below Oregon Caves on trail down Cave Creek.

Distinguishing Features: Occurs in compact and cleavable masses. Sometimes in triclinic crystals similar to augite and diopside in angles. Color, pink or red, often stained black by oxide of manganese. Hardness = 6. Specific gravity = ±3.6. Fuses at (3) to dark colored glass. Partially soluble in HCL.

Occurrence: In crystalline limestones associated with frank-linite, zincite, willemite and calcite. In veins.

Uses: Ornamental stone. Large amount being used in Russia. Some specimens make attractive settings for rings and stick pins.

RHODOCHROSITE MnCO₃

Locality: Josephine County, Waldo district on Cave Creek, three miles below Oregon Caves on trail down Cave Creek.

Distinguishing Features: Occurs in cleavable masses and rhombohedral crystals. Cleavage, rhombohedral. Hardness = 4. Specific gravity = 3.5. Color, pink or reddish-brown. Surface blackens on exposure. Infusible. Borax bead in oxidizing flame is amethyst color. Effervesces in warm HCL. (Distinction from rhodonite.)

Occurrence: In veins. It is the gangue mineral at Butte, Montana. As a secondary mineral formed from rhodonite.

Uses: In some cases an ore of manganese.

ROCK-GYPSUM
(See Gypsum)

RUTILE Ti O_2

Locality: Douglas County, Roseburg quadrangle. (See Roseburg Folio.)

Distinguishing Features: Occurs in tetragonal crystals usually prismatic in habit; also in embedded grains, as acicular inclusions or in massive form. Twins are common. Hardness = 6—6 ½. Specific gravity = ±4.3. Color, red, reddish brown to black. Deep red by transmitted light. Streak pale brown. Crystals often show deep striations or lines on the prism faces. Infusible. Gives violet Na PO_3 bead in reducing flame. Insoluble in acids.

Occurrence: As a secondary mineral in metamorphic rocks, i. e., gneiss, schists, etc. Occurs as a paramorph after brookite. As an accessory constituent of igneous rocks. Sometimes occurs as needle-like crystals in quartz. As an accessory mineral in apatite veins in gabbro.

Use: As a source of ferro-titanium. As a coloring matter for porcelain.

SELENITE
(See Gypsum)

SATIN SPAR
(See Gypsum)

SAPPHIRE Al_2O_3

Locality: Harney County, claims were filed (1902) on sections 1, 6 and 17, T. 25 S., R. 35 E.

Distinguishing Features: Sapphire is the blue transparent variety of corundum. Color, blue and white. Hardness = 9. Specific gravity = —3.95—4-1. Infusible. Heated with cobalt nitrate solution, the mineral becomes deep blue. Insoluble in acids.

Occurrence: In igneous rocks such as syenite and nepheline-syenite. In the borders of peridotites and adjacent rocks. In crystalline limestones. In sands and gravels. The Ceylon gravels furnish sapphires.

Uses: Gems.

SIDERITE Fe CO_3
(Spathic Iron, Brown Spar)

Locality: Josephine County, Galice district. **Grant County,** Sumpter quadrangle.

Distinguishing Features: Occurs in small rhombohedral crystals in cavities, in botryoidal crusts, in cleavable and compact masses; also as concretions in shale. Cleavage, rhombohedral at angles of 73° and 107°. Hardness = 3½—4. Specific gravity = ±3.8. Color, light and dark shades of brown and gray. Difficult to fuse. Becomes magnetic when heated on charcoal. Effervesces in hot HCL.

Occurrence: Siderite occurs as a vein mineral. As concretions in shale. In cavities in basalt as a secondary mineral.

Use: Minor ore of iron.

SILVER Ag
(Placer and Lode)

Locality: **Baker County,** Baker (1), Burnt River (1), Cornucopia (1), Cracker Creek (3), Mormon Basin (2), Pine Creek, Rye Valley (1), Sumpter (2), Unity (1), Weatherby (2) and Whitney (1) districts. **Coos County,** Whiskey Run (1), Flanagan Bar (1) and Johnson Creek (1) districts. **Crook County,** at Howard (1) (in black sand). **Curry County,** Ophir (1), Chetco (1), Gold Beach (1), Elk Creek (1) and Sixes River (1) districts. **Douglas County,** Bohemia (2), Canyonville (1), Cow Creek (3), Green Mountain (1), Myrtle Creek (1), Perdue (1), Poker Flat (1) and Ollala (1). **Grant County,** Beech Creek (1), Bull Run (1), Gold Center (1), Canyon Mountain (1), Elk Creek (1), Granite (1), Poker Flat (1), Quartzburg (2) and Susanville (1) districts. **Jackson County,** Applegate (3), Foots Creek (3), Forest Creek (1) and Jacksonville (3) districts. **Josephine County,** Althouse (1), Applegate (1), Briggs Creek (1), Galice (3), Grave Creek (3), Illinois (1), Jump-Off Joe (3), Louse Creek (1), Pickett Creek (1), Rogue River (1), Sucker Creek (2), Waldo (3), Williams (1), Wolf Creek (1) and Sweat Basin (1) districts. **Lane County,** Bohemia (2) and Blue River (2) districts. **Malheur County,** Glengarry (1), Harper (1), Lutz (1) and Quartz Gulch (1). **Wheeler County,** Spanish Gulch district(1).

(1) Placer. (2) Lode. (3) Placer and lode.

Distinguishing Features: Occurs in wirelike forms, thin sheets, dendritic groups and masses. Very small cubical crystals are rare. Hardness = 2½—3. Specific gravity = ±10.5. Color, white, but tarnishes to dark colors. Malleable and ductile. On charcoal, fuses to a white globule. Soluble in HNO_3 or H_2SO_4 and is precipitated from this solution by HCL. The precipitate white and curdy turns purple on exposure to the sun.

Occurrence: In veins in free state and combined with other minerals.

Uses: The uses of silver are many and are well known.

SPHALERITE Zn S
(Zinc Blende, Black Jack, False Galena)

Locality: **Baker County,** Cable Cove, Greenhorn, Baker, Sparta, Elkhorn and Sumpter districts. **Jackson County,** Upper Applegate and Gold Hill districts. **Grant County,** Quartzburg, Susanville and Alamo districts. **Lane and Douglas County,** Bohemia district. **Lane County,** Blue River mining district. **Linn County,** Santiam. **Josephine County,** Almeda mine, Galice district. **Curry County,** Mt. Emery or Emily.

Distinguishing Features: Occurs in compact cleavable masses. Imperfect inclined hemihedral isometric crystals. Color, yellow, brown. Resinous luster. Streak is lighter than the color of the mineral. Cleavage very good dodecahedral at angle of 60° and 90°. Hardness = 3.5—4. Specific gravity = 3.9—4-1. Difficult to fuse. Sublimate on charcoal is white when hot and yellow when cold. Moisten sublimate with cobalt nitrate and heat intensely, a green color is produced in the sublimate.

Occurrence: In veins associated with various minerals especially galena. Occurs also in sedimentary rock as an accessory mineral.

Use: Ore of zinc.

STIBNITE $Sb_2 S_3$

Locality: **Baker County,** Sumpter and Virtue districts. **Jackson County,** two miles north of Watkins in fractures in greenstone; Ashland district in Barron mine and also 30 miles east of Medford. **Lane and Douglas Counties,** Bohemia district. **Crook County,** Trout Creek mines.

Distinguishing Features: Occurs in columnar or bladed aggregates. Sometimes in acicular crystals and in granular masses. Perfect pinacoidal cleavage yielding bladelike strips. Very brittle. Easily cut with knife. Hardness = 2. Specific gravity = ±4.5. Fuses very easily at (1) on charcoal, giving a dense white sublimate near the assay and a yellow deposit farther away. The fumes given off are dense white with a sulphur odor. Decomposed by HNO_3 with the separation of ($HSbO_3$).

Occurrence: In veins with sphalerite, pyrite, cinnabar, galena and realgar. The gangue is composed of quartz, barite or calcite.

Uses: Source of antimony. Antimony is used chiefly in manufacture of safety matches, fireworks, percussion caps, type metal and other alloys.

SYLVANITE (Au, Ag) Te$_2$

Locality: Baker County, Cornucopia district in quartz veins at Bryan mine. Josephine County, Grants Pass district.

Distinguishing Features: Occurs in monoclinic crystals which are bladed or prismatic and vertically striated. The crystals are indistinct. Sylvanite also occurs in granular masses. Hardness = 1.5—2. Specific gravity = 8. Luster, brilliant metallic. Color, silver white inclining to steel gray. Brittle. Cleavage is good pinacoidal. Fuses easily at (1) on charcoal to a gray bead. Colors flame green and gives a white sublimate. With soda reduces to a yellow button. The powdered mineral dropped in hot concentrated H$_2$SO$_4$ colors the solution purple-red.

Occurrence: As a vein mineral. Mineral occurs in dacite dikes in Transylvania.

Uses: Ore of gold and silver. Boulder, Colorado, West Austraila and Transylvania are type localities.

TALC H$_2$ Mg$_3$ Si$_4$ O$_{12}$
(Talcum)

Locality: Clackamas County, Eagle Creek. Jackson County, near Woodville. Josephine County, Grants Pass.

Distinguishing Features: Occurs in foliated, compact and fibrous masses, also in scales. Distinct crystals are not common. Cleavage is perfect basal. The plates are flexible, not elastic. Hardness = 1. Specific gravity = 2.7. Color, gray, white or pale green. Pearly luster. Greasy or soapy feel. Insoluble in acids.

Occurrence: Talc occurs as a secondary mineral formed by the alteration of silicates, like serpentine, actinolite, etc. In talc-schists and soapstones.

Uses: Soap, talcum powder, French chalk and fibrous material in paper manufacture.

TENORITE Cu O
(Melaconite, Black Copper)

Locality: Josephine County, Queen of Bronze mine, Waldo district.

Distinguishing Features: Occurs earthy and massive. Dull luster. Color, black to iron gray. Black streak. Hardness = 3.4. Specific gravity = —6.0. Infusible in blowpipe flame. Gives copper flame (azure-blue when moistened with HCL). Borax beads are blue in oxidizing flame and opaque red in the reducing flame. Tenorite is a common alteration product of copper minerals occurring as a coating or as an associate.

Occurrence: In oxidized ore bodies as a secondary mineral. (Cuprite is more common.)

TETRAHEDRITE $Cu_8 Sb_2 S_7$
(Gray Copper)

Locality: **Baker County,** Greenhorn, Virtue, Sumpter and Cable Cove districts. **Grant County,** Badger mine in Susanville district.

Distinguishing Features: Occurs in tetrahedrons of the isometric system. Hardness = 3½—4. Specific gravity = 4.8. Color, steel gray. Streak, gray. No cleavage and uneven fracture. Brittle. Fuses easily, giving off dense white fumes and white sublimate near the assay. Gives a green solution with HNO_3 with separation of sulphur and a white residue $HSbO_3$.

NOTE.—Argentiferous tetrahedrite is called Freibergite and is an ore of silver. Schwatzite is mercury bearing tetrahedrite.

Occurrence: In veins as an associate of chalcopyrite.

Uses: Ore of copper called "gray copper" by miners. Tetrahedrite is sometimes silver-bearing and is then an ore of silver.

THOMSONITE $(Ca Na_2)_2 Al_4 (Si O_4)_4.5H_2O$

Locality: **Lane County,** in road cut 100 yards west of Deerhorn; also in gravel of Willamette and McKenzie Rivers near Eugene.

Distinguishing Features: Occurs in aggregates of radiating platy orthorhombic crystals. The crystals radiate from one or more centers. Color, snow-white. Sometimes reddish, yellowish or greenish. Cleavage, brachy-pinacoidal. Fuses easily with intumescence to white enamel.

Occurrence: In cavities in igneous rocks. As pebbles in stream gravels.

Use: Some pieces are polished and used for ornamental purposes.

WAD Mn $O_2 + H_2O$ (impure)
(Bog Manganese)

Locality: Baker County, Sumpter quadrangle.

Distinguishing Features: Wad, or bog manganese, consists chiefly of oxide of manganese and water with some oxide of iron, silica, alumina, copper, cobalt, lithium, or barium. Occurs in rounded porous masses of a brownish color. Streak, brown. Dull luster. Infusible. Gives water in a closed tube. Gives manganese bead tests. (See test under Pyrolusite.) Soluble in H Cl, chlorine is given off.

Occurrence: In bog deposits at times associated with limonite. As secondary mineral formed by the alteration of manganese minerals.

Uses: Used as a pigment and in the manufacture of chlorine. Cobalt-bearing wad from New Caledonia is a source of cobalt.

ZINC
(See Sphalerite)

ZIRCON Zr Si O_4

Locality: **Baker County,** Upper Burnt River in black sands and Bonanza placers. **Curry County,** Sixes River in placer gravel. **Coos County,** in elevated beach gravels between Three Mile Creek and The Lagoons. **Baker County,** Durkee, Anthony, Sumpter and Rye Valley (in black sand). **Clatsop County,** Astoria, Clatsop Beach, Warrenton, Hammond, Fort Stevens, Carnahan Station, Seaside, Clatsop, Morrison, Clatsop Spit and Elk Creek (in black sand). **Coos County,** Marshfield, Bullards, South Fork Coquille River, Randolph district (old and present beaches), Whiskey Run, Whiskey River, Bandon Beach and Johnson Gulch (in black sand). **Curry County,** Gold Beach, Chetco, Port Orford, Rogue River beach, near Pistol River and Cuneffs Beach (in black sand). **Douglas County,** South Umpqua River and Riddle (in black sand). **Grant County,** Granite (in black sand). **Jackson County,** Ashland, Weimer, Gold Hill, Jacksonville (in black sand). **Josephine County,** Josephine Creek, near Kerby, Holland, Kerby, Sutler Creek, Waldo, Sucker Creek, Wolf Creek, Placer and Greenback (in black sand). **Lincoln County,** Coos Bay, Yaquina Bay and Newport (in black sand). **Linn County,** Foster (in black sand). **Malheur County,** Snake River (in black sand). **Multnomah County,** Fulton, Latourell (in black sand). **Polk County,** Falls City (in black sand). **Umatilla County,** Weston (in black sand). **Wasco County,** Hood River (in black sand).

Distinguishing Features: Occurs as tetragonal crystals in sands, as loose crystals, or imbedded. The crystals are usually small and made up of a flat pyramid and prism. Hardness = 7. Specific gravity = ±4.6. Color, generally brown. When transparent, appears red. Brittle. Infusible, but loses color easily. Insoluble in acids.

Occurrence: In sands and gravels. In igneous rocks especially syenite and soda-granites.

ZOISITE H Ca_2 Al_3 $(Si O_4)_3$

Locality: Douglas County, Roseburg quadrangle near Roseburg. (See Roseburg folio.)

Distinguishing Features: Occurs usually in columnar masses. Orthorhombic crystals are rare. Cleavage perfect—parallel to side (OlO). Hardness = 6—6½. Specific gravity = ±3.4. Color usually gray, sometimes pink. Fuses with swelling to a white glass (fusion not easily made). Gives water in close tube. Gelatinizes in HCl after the mineral has been powdered and fused.

Occurrence: Occurs generally in Hornblende and glaucophane schists. (This is the case with the Roseburg specimens.)

GLOSSARY

A

Accessory mineral. Any mineral not regarded as an essential constituent of a rock, although it may be frequently present.

Acicular. Needle-like.

Adamantine luster. Luster of diamond.

Ag Cl. Silver chloride.

Aggregates. Groups or bunches.

Amalgam of gold. Alloy of gold and quicksilver.

Amorphous. Substance without crystal form or structure like glass.

Amygdule. Rounded nodules of secondary mineral matter filling cavities in rocks.

Amygdaloidal. See Amygdule. (A rock which has numerous amygdules is said to have an amygdaloidal structure.)

Aqua Regia. A mixture of three parts of concentrated HCl and one part of concentrated HNO_3.

B

Basic (igneous) rock. An igneous rock comparatively low in silica.

Botryoidal. A form resembling a bunch of grapes.

C

Calcarous tufa. A porous deposit of calcium carbonate formed by hot springs.

Cleavage. Property possessed by some minerals which enables them to break along certain lines of weakness; i. e., parallel to crystal faces or possible crystal faces.

Clino-pinacoid. A form parallel to the (a) and perpendicular to the (b) axis in the monoclinic system.

Columnar. Made up of column-like masses.

Conchoidal fracture. Fracture producing a surface covered with concentric curves like the inside surface of a shell.

Concretions. Rounded nodules in sedimentary rocks, especially shales and limestones.

Contact mineral. Mineral formed in the contact zone.

Contact-zone. The zone bordering the contact of igneous and sedimentary rocks.

Crystal. A mineral bounded wholly or partially by plane surfaces called faces.

D

Dacite.   Rock composed chiefly of lime-soda-feldspars **and** quartz.

Deflagrates.   Showing sudden combustion.

Dentritic. Branching, tree-like.   A form taken by some minerals like native copper.

Dike.   Intruded rock, filling cracks in the earth's crust. Dikes stand perpendicular or nearly so.

Disseminated.   Scattered through something.

Ductile.   Capable of being drawn into wire.

E

Effervesces.   The bubbling caused by the action of an acid on a carbonate.

Elastic.   Having the power to spring back to its original position after the disturbing force has been removed.

Exfoliation.   Swelling or spreading out of a mineral when heated.   When used in speaking of rocks it means the shelling off in concentric layers, due to the unequal expansion and contraction of the rock surface.

F

Fibrous.   Made up of fibers.

Flexible.   Capable of being bent, but will not regain its former position when the disturbing force has been removed.

Foliated.   Made up of plates.

Form.   A form in crystallography includes all the faces which have a similar position with relation to the planes or axes of symmetry.

Fusible.   Capable of being melted in a flame.

G

Gangue mineral.   The valueless minerals, associated with ore minerals.

Gelatinization.   Formation of a jelly-like substance when the powdered mineral is boiled in acids.

Geodes.   Rounded nodules with hollow centers which are lined with crystals chiefly quartz and calcite.

Gneiss.   A metamorphosed (changed) rock, composed chiefly of orthoclase feldspar, quartz and mica, having a banded structure.

Granite.   An even-grained holocrystalline rock composed chiefly of quartz, orthoclase feldspar, mica and hornblende.

H

Hardness. Refers to the ability of one mineral to scratch another. The hardness of a mineral is stated in terms of Moh's scale of hardness, which is as follows: (1) talc, (2) gypsum, (3) calcite, (4) fluorite, (5) apatite, (6) feldspar, (7) quartz, (8) topaz, (9) corundum, (10) diamond.

H Cl. Hydrochloric acid; also known as muriatic acid.

H NO_3. Nitric acid.

H_2SO_4. Sulphuric acid.

I

Incrusting. Occurring in a thin layer coating some other substance.

Insoluble. Is not dissolved by acids or other solutions.

Intumescence. Bubbling of minerals when heated, due to escaping water.

L

Lamellae. Thin sheets or layers.

Lamprophyrs. Rocks composed chiefly of ferromagnesian minerals, iron ores, pyroxene, hornblende, biotite, olivine and varying amounts of feldspars. These rocks are commonly porphyritic, having a dense dark ground mass.

Long ton = 2,240 pounds.

Luster. The quality of light reflected from the surface of substances.

M

Magmatic segregation. Deposits of ores or minerals formed by segregation from igneous magmas.

Magnetic. Being attracted by a magnet.

Malleable. Capable of being hammered into thin sheets.

Mammillary. A form resembling masses of flattened domes.

Marl. Loose, earthy deposits of intermingled carbonate of lime with clay in variable proportions.

Metallic luster. Luster of metals.

Metasomatic processes. Processes by which minerals replace rocks and minerals, due to the action of mineral-bearing solution.

Micaceous. Composed of thin plates easily separated.

Monoclinic. Refers to the crystal system which includes all crystals which can be referred to three non-interchangeable axes of reference. Two of these axes (a and c) are inclined to each other and (b) is at right angles to (a and c).

N

Nepheline-syenite. Syenite with the mineral nepheline present.

Nitric acid. H NO₃.

NH₄ OH. Ammonium Hydroxide.

Non-crystalline. Occurring without crystal form.

O

Oolitic. Composed of minute spheres about the size of a fish roe . (Like the Cinton, Alabama, oolitic iron ore.)

Orthopinacoid. Form (100) parallel to the (b) axis in the monoclinic system.

Orthorhombic. A crystal system in which we have three crystal axes at right angles and none of equal length.

Oxidized zone. A portion of an ore vein (the part nearest the surface) where the minerals have been oxidized and the vein matter leached.

Oxidizing flame. The extreme outer tip of a flame produced by a blowpipe or Bunsen burner. Most favorable for oxidation.

P

Parting. Ability of some crystals to break along certain planes of weakness due to some molecular distrubance such as twining.

Paramorph. Pseudomorph of one dimorphous mineral after another.

Pegmatite. Rock composed chiefly of orthoclase feldspar and quartz in very large crystals. This is a dike rock occurring in granites and metamorphic rocks, like schist, gneiss, etc.

Peridotites. Igneous rock composed chiefly of olivine and pyroxene.

Pigment. Coloring matter.

Pitchy-luster. Having the luster of pitch.

Pisolitic. Made up of spheres a little larger than oolites.

Polysynthetic twining. Twining which is produced by revolving alternate sections of the crystal 180° about parallel twining planes.

Precipitate. To separate as a precipitate.

Prismatic. Elongated in one direction, but not applying to bladed, needle-like or thread-like crystals.

Pseudohexagonal. Minerals belonging to some system other than the hexagonal will sometimes replace the original hexagonal mineral, taking its form.

Pseudomorphs. A mineral taking the crystal form of another mineral.

Q
Quartzite. Metamorphic rock composed chiefly of quartz.

R
Reducing flame. The tip of the inner luminous cone of the Bunsen burner or blowpipe flame.

Residual clay. Clay found overlying the rock from which it was formed.

Resinous luster. Luster of resin.

Rhombohedron. A class of crystals made up of six faces and resembling a distorted cube. Belongs to the hexagonal system.

Rhyolites. Volcanic igneous rocks composed chiefly of quartz and orthoclase with some glassy material.

S
Sandstone. Sedimentary rock composed chiefly of quartz grains cemented by siliceous, calcareous or argillaceous cement.

Schist. A meta-morphosed (changed) finely laminated rock containing muscovite, biotite, hornblende, chlorite, actinolite, talc, quartz and garnet. Schists are named according to the most prominent mineral; i. e., talc, schist, mica-schist, hornblende-schist, etc.

Short ton = 2,000 pounds.

Secondary mineral. One formed subsequent to the formation of the minerals of the main rock mass.

Sectile. Can be cut with a knife.

Shale. Compact clays and muds possessing a more or less thinly laminated structure.

Silky luster. Luster of silk.

Specific gravity is the weight of a substance compared with the weight of an equal volume of water at 4° centigrade.

Stalactitic. Icicle-like form in which some minerals occur.

Stalagmite. An icicle-like formation of mineral matter forming at the bottom of caves or cavities. (Stalactites build down from the roof and stalagmites build up from the floors of caves.)

Streak. Color of the powdered mineral. Determined by rubbing the mineral to be tested on unglazed porcelain or by scratching with a knife until a powder has been produced.

Striations. Fine lines appearing on the crystal or cleavage face due to polysynthetic twining or other causes.

Sublimate. A deposit formed on charcoal near the assay while the mineral is being fused.

Syenite. Igneous rock composed chiefly of orthoclase feldspars with some hornblende, mica or pyroxene.

T

Tabular. Flat like a table top.

Trachyte. Volcanic igneous rock composed chiefly of orthoclase and little or no quartz.

Translucent. A substance is translucent when light will pass through, but objects cannot be seen through it.

Travertine. A banded calcium carbonate deposit formed by water of springs or rivers.

Twin-crystals. Twin-crystals are those in which one or more parts, regularly arranged, are reversed with respect to the other parts.

V

Variegated. Having marks or patches of different colors.

Veins. Cracks in the earth's crust which have been filled with minerals.

Volatile. Easily vaporized.

Volcanic neck. Hardened lava in the throat of an extinct volcano.

Vitreous luster. Luster of broken glass.

REFERENCES

1. Useful Minerals of the U. S. A.   Bulletin U. S. G. S. No. 585.
   Samuel Sanford and R. W. Stone.

2. Mineral Resources and Mineral Industries of Oregon.   Bulletin University of Oregon, Vol. I, No. 4. New Series.   1903.
   O. F. Stafford.

3. Mineral Resources of Southwestern Oregon.   Bulletin U. S. G. S. No. 546.   J. S. Diller.

4. Mineral Resources of Oregon.   Bulletin Oregon Bureau of Mines and Geology, Vol. I, No. 5.   1914.   A. N. Winchell.

5. U. S. G. S.   Folios Nos. 49, 73, 89 and 103.

6. Dana's System of Mineralogy.

7. Rogers' Study of Minerals.

8. Moses and Parsons' Mineralogy, Crystallography and Blowpipe Analysis.

9. Butler's Handbook of Minerals.

10. Mineral Resources of Oregon.   Bulletin Oregon Bureau of Mines and Geology, Vol. I, No. 6.   1914.   J. T. Pardee and D. F. Hewett, U. S. Grant and G. H. Cady

11. Black Sands of the Pacific Coast.   Publication U. S. G. S. 1907.   David T. Day and R. H. Richards.

12. The Economic Geology Resources of Oregon.   Bulletin Oregon State Bureau of Mines.   1912.   H. M. Parks.

13. American Journal Science, 4th series.   Vol. 6, pp. 66-8. A. S. Eakle.   (Reference to Erionite.)

14. Mineral Resources U. S. pp. 547-550.   1891.

15. University of Oregon Bulletin No. 4, Vol. 10.   New Series. December, 1912.   Bibliography of the Geology, Paleontology, Mineralogy, Petrology and Mineral Resources of Oregon.   Chas. W. Henderson and J. W. Winstanley.

16. Pocket Dictionary of Common Rocks and Rock Minerals. Collier Cobb, Chapel Hill, N. C.

17. Mineral Resources of Oregon, Vol. I, No. 3.   A. J. Collier.

18. Mineral Resources and Geology of Curry County, Oregon, (to be published January, 1916.)   G. Montague Butler, Graham John Mitchell.   Bulletin Oregon Bureau Mines and Geology.

VALUE OF MINERAL PRODUCTS OF THE UNITED STATES
(For years 1904 to 1915.)**

The value of mineral products varies from year to year, but the following figures will give the prospector an idea of the approximate value of his discovery:

Metals

PRODUCT	Average value for years from 1904-1915
Aluminum (98%), per pound	$ .21- .59
Antimony, per pound	.14- .35
Copper, per pound	.18
Gold, per ounce, troy	20.00
Lead, per pound	.04+
Magnesium, per pound	1.50
Mercury, per flask, 75 pounds	50.00-103.00
Nickel, per pound	.50
Platinum, per ounce, troy	44.00-90.00
Pig Iron, per long ton, 2,240 pounds	14.00
Silver, per ounce, troy	.53
Tin, per pound	.32
Zinc, per pound	.07- .14

Non-Metals

Structural Materials:

Kaolinite, per ton	$ 8.00-18.00
Calcite, per ton (burned for lime)	9.75
Gypsum, per ton (calcined)	3.40

Abrasive Materials:

Corundum, per pound	.09
Garnet, per ton	30.00
Diatomaceous earth, per short ton (2,000 pounds)	60.00
Quartz, ground, per ton	11.00
Quartz, lump, per ton	5.00

Chemical Materials:

Borax, per pound	.02
Pyrite, per long ton (2,240 pounds)	3.70
Sulphur, per long ton	17.00
Salt, lump, per ton	4.25
Fluorite (Fluorspar), per short ton (2,000 pounds)	6.40

Pigments:

Barite, crude, per short ton	3.60

Non-Metals—Continued

PRODUCT	Average value for years from 1904-1915
Miscellaneous:	
Asbestos, per short ton	10.00
Asphaltum, per short ton	10.00
Bauxite, per long ton (chief ore of aluminum)	4.60
Chromic iron ore, per long ton	11.00
Graphite, crystalline, per pound	.05
Graphite, amorphous, per short ton	17.00
Garnierite (50%), per ton	15.00
Gypsum, per ton (ground for land-plaster or fertilizer)	1.90
Magnesite, per short ton (2,000 pounds)	8.00
Molybdenite (92%), per pound	.30
Monazite sand (carrying 5% thorium oxide), per pound	.10
Mica, sheet, per pound	.20
Mica, scrap, per short ton	15.00
Pyrolusite (80%-85% MnO_2), per ton	25.00
Talc, per short ton	9.50
Tungsten (50% WO_3), per unit	7.50-45.00
Iron Ores (Value at the Mine):	
Hematite, per long ton	2.21
Brown ore (limonite), per long ton	1.83
Magnetite, per long ton	2.12
Carbonate (siderite), per long ton	2.06

NOTE.—These values were computed from the figures in Mineral Resources of the United States from 1904-1913. Recent variations in the prices of minerals, especially the metals, have been taken from "Mining Press" and "Engineering and Mining Journal."

For those desiring to purchase mineral specimens, the Foote Mineral Company of 107 N 19th Street, Philadelphia, Pa., is recommended. This company will furnish upon request a catalogue with prices of minerals.